Discovering
Advanced
Algebra
An Investigative Approach

SECOND EDITION

Calculator Notes for the
Texas Instruments TI-83 Plus and TI-84 Plus

DISCOVERING

MATHEMATICS
™

Key Curriculum Press
Innovators in Mathematics Education

Editor: Heather Dever

Project Administrator: Tamar Wolins

Writers: Larry Copes, Eric Kamischke, David Rasmussen

Accuracy Checker: Nicole Nagappan

Production Editor: Christa Edwards

Editorial Production Supervisor: Kristin Ferraioli

Production Director: Christine Osborne

Senior Production Coordinator: Ann Rothenbuhler

Text Designer: Jenny Somerville

Composition, Technical Art: ICC Macmillan Inc.

Cover Designers: Jill Kongabel, Jeff Williams

Printer: Versa Press, Inc.

Textbook Product Manager: Tim Pope

Executive Editor: Josephine Noah

Publisher: Steven Rasmussen

Cover Photo Credits: Background and center images: NASA;
all other images: Ken Karp Photography.

Key Curriculum Press
1150 65th Street
Emeryville, CA 94608
(510) 595-7000
editorial@keypress.com
www.keypress.com

Printed in the United States of America
10 9 8 7 6 5 4 3 2 1   13 12 11 10 09   ISBN 978-1-60440-013-7

Contents

Chapter 11

Chapter 12

Chapter 13

Introduction

To accommodate students with different—and ever-changing—types of graphing calculators, in *Discovering Advanced Algebra* we refer to calculators generically in the student book and provide detailed notes for many different types of calculators in separate books. As new calculator technology is introduced and gains acceptance in secondary mathematics classrooms, Key Curriculum Press will create new calculator notes to accommodate the changing technology.

This book is written for use with the Texas Instruments TI-83 Plus and TI-84 Plus graphing calculators. The notes are designed to familiarize students with calculator use and to provide specific keystroke instructions. Some notes help students use motion sensors, such as the Texas Instruments Calculator-Based Ranger (CBR2) or Calculator-Based Laboratory (CBL2), to collect data with their calculators. Other notes contain programs for specific investigations or exercises.

Your students will find references to the calculator notes throughout the student book. For example, on page 191, you will find the reference [▶☐ See **Calculator Note 4A** to learn about defining and evaluating functions. ◀]. This reference indicates that in Calculator Note 4A there are instructions on how to use the calculator to evaluate functions. All pertinent calculator notes for each lesson are also mentioned in the materials list of the *Discovering Advanced Algebra Teacher's Edition*. How much your students need these notes will depend on their experience with graphing calculators and with the particular graphing calculator methods used to explore concepts in *Discovering Advanced Algebra*. The notes will be particularly useful if your students use many different types of calculators.

You may want to copy and distribute the notes as they are needed, or you can copy and distribute all the notes for each chapter as you begin work on that chapter. You can choose a strategy based on your students' specific needs, your access to a copy machine, and your duplicating budget. If your students have had limited experience with graphing calculators, an ideal strategy is to distribute a copy of the notes to each student and encourage students to keep the notes in a special section of their notebooks. Another strategy is to make enough copies for each group of students to have access to one or two copies of the notes, stored in either three-hole report covers or individually in hanging files. If your students have had a lot of experience with graphing calculators, however, you may need only one or two copies of the notes for classroom reference. Place the copies in binders and make them available for students to check out. If your students use many different calculator brands and models, you'll need to make copies of the notes for each type of calculator.

If students need home access to a note, they will find all the notes on the *Discovering Advanced Algebra* website, *www.keymath.com/DAA*.

Even if you don't usually copy a complete set of calculator notes for each student, you may find it helpful to distribute copies to all students for particular sections of material. For example, some of the sections in the student book contain special calculator programs. If students manually input these programs rather than link them, they may need access to a hard copy of the program. For shorter programs you can display the program commands with an overhead projector, but if students are using a variety of calculators, you'll probably be better off providing each student with notes for his or her particular calculator. If you have TI Connect linking software and access to a computer, you can take advantage of the programs and data stored on the *Teaching Resources* CD available with the

Teaching Resources package and also at *www.keymath.com/DAA* (for students) or *www.keypress.com/keyonline* (for teachers). You can download programs or data from the CD or Key's websites to a computer and then to a calculator linked to the computer. Students can link their calculators to the computer or to other calculators to transfer the data and programs. By downloading programs and data in this way, you and your students can avoid the hassle of debugging programs and checking the accuracy of data input.

You will also find it useful to make available to students a copy of the *TI-83 Plus Graphing Calculator Guidebook* or the *TI-84 Plus Graphing Calculator Guidebook*. See *http://education.ti.com* to learn about further resources Texas Instruments provides to teachers.

Handheld Software Applications

Many Texas Instruments calculators—including the TI-83 Plus and TI-84 Plus— allow you to add Handheld Software Applications (Apps) that increase the calculator's functionality, in much the same way that you add software to a computer. An App is different from a program because you can neither edit nor modify an App. Also, an App is often registered to a specific calculator, so you may not be able to share it by linking calculators.

A variety of sources provide Apps, and new Apps are released frequently. Some Apps are free while others require that you purchase them. Furthermore, not every App is available for every model of calculator. Because Apps are gaining popularity, it is important to recognize which Apps are truly beneficial and the best way to incorporate them into *Discovering Advanced Algebra.*

For this edition, we have decided to provide notes for only free Apps developed by Texas Instruments that provide improved functionality within the context of *Discovering Advanced Algebra.* These Apps are available on the *Calculator Data and Programs* CD. (Please read the licensing agreement for information about using these Apps within your classroom.)

We also recognize that you may find other Apps beneficial. You may want to search the Internet for other Apps that could be worthwhile for your classroom.

The TI-83 Plus and TI-84 Plus Data App

The *Calculator Data and Programs* CD includes a TI-83/84 Plus App, DiscAdAl.8xk, that loads and archives select data files from *Discovering Advanced Algebra.* (The CD also contains a text file called **Data TOC,** which tells how the data files correlate to the textbook.) This App does not require registration and you can load it free of charge onto any student's TI-83 Plus or TI-84 Plus. After loading the App, press APPS and select the application DiscAdAl. You can then select the appropriate chapter and lesson, press enter to select the data list(s) (the list name will have a box to the left when it is selected), and then load the data list(s). The data lists will now appear under the NAMES submenu when you press 2nd [LIST]. To use the data in the Stat Edit screen, you have three options: select SetUpEditor when loading the lists to automatically put them in the Stat Edit screen; define one of the standard lists as equivalent to the data list (arrow up to a list name, press 2nd [LIST], and select the name of the data list); or add a new column for the list (arrow up to a list name and arrow right until you are given a new column with no name, then press 2nd [LIST] and select the name of the data list).

Getting Started • Navigating Screens and Menus

You'll use a variety of screens while working with the calculator. These are the ones you'll use most often.

Home Screen

Press 2nd [QUIT]. This screen usually comes up when you turn on the calculator. You'll do almost all your calculations here.

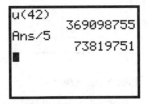

Mode Screen

Press MODE to change the number of decimal places displayed, the style of graph displayed, and other settings as necessary. Most of the time, your Mode screen should look like this one.

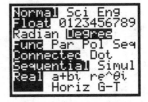

Graph Screen

Press GRAPH to display graphs.

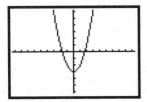

Window Screen

Press WINDOW to set the window of values that you want to graph.

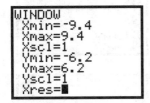

Y= Screen

Press Y= to enter equations that you want to graph or evaluate.

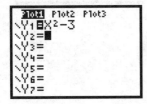

(continued)

Stat Edit Screen

Press STAT ENTER to enter and work with lists.

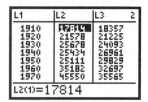

Plot Setup Screen

Press 2nd [STAT PLOT] to set up a box plot, histogram, or other statistics plot.

Commands

There are keys for the most common commands you'll use—numbers and operations, for example, and DEL (delete) and 2nd [INS] (insert). You'll choose other commands from menus and submenus. For example, press MATH and you will see four submenus: MATH, NUMber, ComPleX, and PRoBability. Use the right and left arrow keys to move among submenus. With each submenu, there is a list of commands. Use the up and down arrow keys followed by ENTER, or type a number, to select one of the commands. For example, to select the lcm(command in the NUM submenu, arrow right to NUM and then either arrow down repeatedly, or up twice, and press ENTER. You'll return to the Home screen. If you now type two integers, such as 18 and 24, separated by a comma, close the parentheses, and press ENTER, the calculator will display the least common multiple of 18 and 24, which is 72.

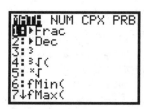

 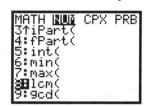 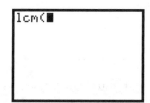 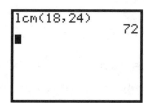

Finding a Command

To find a command, you can press 2nd [CATALOG] and then the first letter of the command (letters are printed in green on the calculator surface above the keys). Then use the arrow keys to scroll to and select the command. (On some calculators you can then press + to recall the parameters of the command.)

Note 0A • Fractions and Decimals

To convert fractions to decimals, use the division operation. (For example, to convert 3/5, press [3] [÷] [5].) You can also convert many decimals to fractions and reduce them to lowest terms. Type in the decimal you want, and press [MATH] [ENTER]. The window will show ►Frac and pressing [ENTER] again will give you the fraction. Or, if you've already obtained a decimal result of a calculation, press [MATH] [ENTER] [ENTER] to see the result as a fraction in lowest terms. If the calculator does nothing to the number, the denominator (in reduced form) is more than 10,000.

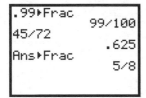
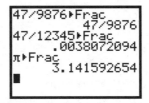

Note 0B • Order of Operations

To evaluate expressions, the calculator uses the standard order of operations, PEMDAS (parentheses, exponents, multiplication and division, addition and subtraction). For example, when you enter the expression 1+12/4*3

a. The calculator reads no parentheses or exponents.

b. The calculator does multiplication and division from left to right: it evaluates 12/4 as 3 to get 1 + 3 * 3, and it evaluates 3 * 3 as 9 to get 1 + 9.

c. The calculator does addition and subtraction from left to right: it evaluates 1 + 9 as 10.

Negation is on the same level as multiplication, so powers are calculated before negation. For example, -4^2 evaluates as -16. To square -4, you use parentheses: $(-4)^2$.

Although there are keys for the brackets and braces you use for grouping when working on paper, the calculator uses them for other things. To group on the calculator, you use only the parentheses keys. So, to evaluate the expression $\frac{5+3}{2 \cdot 4} + \frac{\sqrt{8+1}-4}{(2(7-2))^2}$, you would enter (5+3)/(2*4)+ ($\sqrt{}$(8+1)−4)/(2(7−2))². Study these examples and verify the results on your calculator.

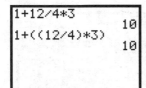
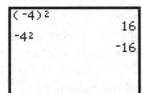
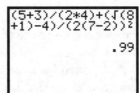

Note 1A • Reentry

If you want to do further calculation on a result you've just found, and that result is the first number in the expression you now want to evaluate, you can simply continue the expression. For example, if you've calculated a result of 3.647483, and you want to multiply by 16, press ⊠ ① ⑥ [ENTER]. If, on the other hand, you want to take the square root of that number, press [2nd] [√] and then [2nd] [ANS] [ENTER] to calculate $\sqrt{3.647483}$.

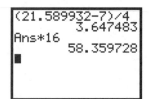

You can also recall and edit a previous expression. Press [2nd] [ENTRY] and use the arrows to move across the expression and enter replacement characters. To delete characters press [DEL], and to insert new characters press [2nd] [INS]. When you are finished, press [ENTER] to recalculate the revised expression.

Repeatedly pressing [2nd] [ENTRY] takes you back to previously evaluated expressions. The number of expressions you can recall depends on their length.

Note 1B • Recursion

The command [2nd] [ANS] allows you to use the result of your last calculation in your next calculation. Also, if you press [ENTER] without pressing another key, the calculator will recompute the last expression. Using these two commands together gives you a recursion machine.

Start by entering the starting value of a sequence. Press [ENTER]. Now enter the rule, using [2nd] [ANS] in place of u_{n-1}. Press [ENTER] repeatedly to generate the sequence. For example, this screen shows

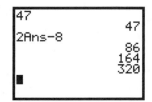

$u_1 = 47$
$u_n = 2u_{n-1} - 8$ where $n \geq 2$

If you go too far in the sequence, you cannot back up. You must start the process over by entering the starting value again, then the rule. You also have to start over if you lose count of the number of terms in your sequence.

One way to avoid losing count of your terms is to generate two recursive sequences at once, the first sequence counting the terms of the second. Use braces, { and }, to enclose the two sequences. This example shows the starting values of 1 and 47 in braces and separated by a comma. Use Ans(1) and Ans(2) in the rules to refer to the previous values. Here Ans(2) does not mean Ans times 2, but rather the second value of the previous list. You still have to start over if you go too far, but you can keep track of how many terms you've generated.

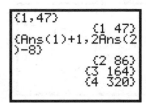

(continued)

You can also use list recursion to keep track of more than two sequences at a time. In these screens the recursive formulas are

$u_1 = 500$

$u_n = (1 + 0.07)u_{n-1}$ where $n \geq 2$ and $u_1 = 500$

$u_n = (1 + 0.085)u_{n-1}$ where $n \geq 2$

If the answer list is too long, scroll to the right with the arrow key to see the last value(s). You can keep answers to a fixed length using a setting on the Mode screen that specifies the number of decimal places displayed. (See **Getting Started** for instructions on moving to and from the Mode screen.)

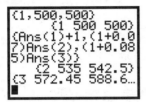 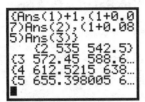 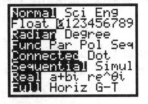 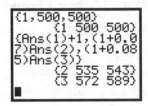

Note 1C/App • Making Spreadsheets Using the CellSheet App

You can use the CellSheet application from Texas Instruments to organize and calculate your table values. Press APPS and select CellSheet. Press any key twice to get to the spreadsheet.

To enter values, arrow to the cell you want to enter into, type the value, and press ENTER. If you want to enter a formula, press STO→ to get the equal sign, then type the formula and press ENTER. To refer to another cell in the spreadsheet, use a letter for its column, and a number for its row. For example, the first cell in the second column is cell B1. (Press ALPHA to enter letters.)

You can enter values for a range of cells at one time: Press Menu (GRAPH), select 3:Options, then select 2:Fill Range. Type your range and formula, then press ENTER ENTER to display the results.

Range: Enter the first and last cells you want to fill, using a colon in between. (If you want to fill more than one column, the first cell is the top-left one, and the last is the bottom-right one.)

(continued)

Formula: Press [STO→] to get an equal sign, then enter the formula for the first cell to be filled. The calculator will automatically adjust the cell references for the other cells to be filled.

   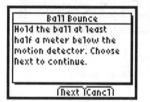

Note 1D/App • Looking for the Rebound Using the EasyData App

You will need a CBR2 (Calculator-Based Ranger).

Connect the CBR2 to the calculator. Press [APPS] and select EasyData. The CBR2 will immediately begin collecting distance data, which is displayed on your calculator screen.

To collect ball bounce data, press Setup ([WINDOW]) and select 5:Ball Bounce. Press Start ([ZOOM]), then Next ([ZOOM] again). If you want to disconnect the CBR2 while collecting the data, do so now. On the CBR2, press [TRIGGER] or [START/STOP] to begin collecting data. The CBR2 will collect data for 5 seconds, or you can press the trigger to stop it sooner. If you didn't get good data, press the trigger to start again. When you have finished, reconnect the CBR2 to the calculator and press Next ([ZOOM]). The calculator will display the data in a graph. You can trace this graph using the left and right arrow keys.

 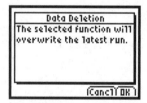

If you need to redo the experiment, press Main ([TRACE]), then press Start ([ZOOM]). You will be told that this function will override the previous data. Press OK ([GRAPH]) and repeat the steps above.

To end the application, press Main ([TRACE]), then press Quit ([GRAPH]). You will get a message telling you where the data are stored. Time data are in L_1, distance data are in L_6, velocity data are in L_7, and acceleration data are in L_8.

Note 1E • Entering Data

The calculator keeps track of data through lists. It has six standard lists, lists L_1 through L_6. To refer to these lists, press [2nd] [L_1] through [2nd] [L_6].

(continued)

There are several ways to enter data into a list. No matter how you enter the data, you can plot and trace the data using instructions from **Notes 1F** and **1G.**

Clearing Data

If a list already has data in it, arrow up to the list name and press CLEAR ENTER.

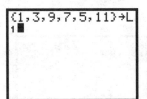

Entering Data into a List from the Home Screen

If you are working with a short list, you may want to enter it from the Home screen. If you enter 2nd [{] 1, 3, 9, 7, 5, 11 2nd [}] STO→ 2nd [L1] ENTER, list L1 will contain those six numbers. To view the list on the Home screen, press 2nd [L1] ENTER.

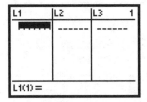

You can also enter a list into the Home screen without storing it in a stat list.

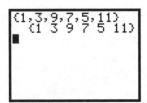

Entering Data Directly into a List

To enter a list into the Stat Edit screen, press STAT ENTER. You'll see three lists. You can arrow to the left or right to see the other three lists. (If the six standard lists don't appear, press STAT, select EDIT, arrow down to 5:SetUpEditor, then press ENTER.)

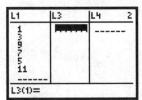
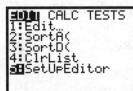
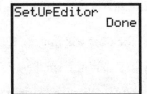

(continued)

Enter or edit values in the list by typing numbers, expressions, fractions, or functions. Press ENTER after each value. All values are converted to decimals.

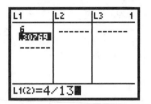 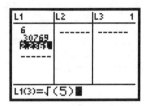 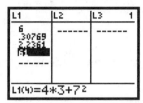

Note 1F • Plotting Data

To set up a plot of the data stored in a list, press 2nd [STAT PLOT] and select one of the plots by scrolling down and pressing ENTER. Then follow these steps:

    a. Select On.

    b. Select one of the six plot forms: scatter plot, xyline plot, histogram, modified box plot, regular box plot, or normal probability plot (not used in this course).

    c. Enter the lists to be used in the stat plot. For one-variable plots (box plots and histograms) enter one list, but for scatter plots and xyline plots enter a list into Xlist for the x-axis and a list into Ylist for the y-axis.

 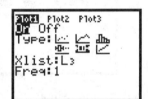  

    d. For one-variable plots, Frequency indicates the number of times each data point occurs in the data set. Usually Freq is set as 1.

    e. For scatter plots, xyline plots, and modified box plots, select the Mark to use in the plot. If you graph more than one plot at the same time, use a different Mark for each plot.

Before viewing the plot, you need to decide what part of the graph you want to view. Press WINDOW.

    Xmin = a number slightly less than the smallest x-value you want displayed.

    Xmax = a number slightly greater than the largest x-value you want displayed.

    Xscl = and Yscl = the distance between tick marks on the two axes. The number of divisions should be less than 25. If there are too many tick marks, the axes will appear too thick.

    Ymin = a number slightly less than the smallest y-value.

    Ymax = a number slightly greater than the largest y-value.

    Xres = 1.

(continued)

For more information about setting the window for box plots and histograms, see **Notes 2B** and **2C.**

Press GRAPH to see the plot.

Note 1G • Tracing Data Plots

If you have a plot displayed and you press TRACE, a "spider" will appear on the plot. Use the right and left arrow keys to move the spider along the plot. The spider's position is given at the bottom of the screen.

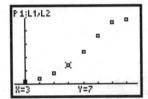 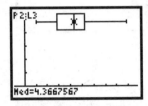 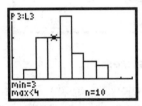

For scatter plots the data are traced in the order they appear in the list, so pressing the left arrow may not move the spider left. One-variable plots always trace the same way, histograms from the left and box plots from the center.

If you have displayed several plots at once, the spider will begin on the first stat plot that is turned on. Pressing the up and down arrow keys makes the spider jump to another stat plot. The top of the screen tells you the plot the spider is on and the lists being used.

Note 1H • Sharing Data

You can copy lists from one calculator into another. This can save time and ensure that you're working from the same data set as others. You will need a link cable and two compatible calculators. The TI-83, TI-83 Plus, TI-84 Plus, and TI-Nspire in TI-84 Plus mode can all share list information.

Push the plug firmly into the ports at the bases of both calculators. Press 2nd [LINK] on each calculator. On the receiving calculator, choose RECEIVE and press ENTER. This calculator should read Waiting… at the top of the screen.

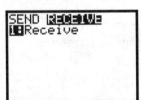 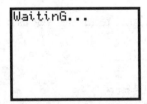

On the calculator with the data, select 4:List…. Arrow down to a list you want to send and press ENTER. This marks the list but does not send it. Mark each list you wish to send.

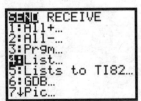 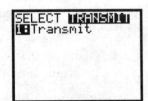

(continued)

When you have marked all the lists, press the right or left arrow to go to the TRANSMIT submenu. Press ENTER. If either calculator gives a LINK ERROR message, then push the link cable in again and start over. If the list you are sending already exists in the receiving calculator, choose 2:Overwrite to replace this list with the new list.

Note 1I • Creating Sequences

Sequence mode is a powerful way of working with recursive formulas. Press MODE, scroll down to the fourth line, and select Seq. Then go to the Y= screen.

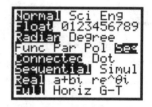

Follow these steps to enter the recursive formula

$u_1 = 47$
$u_n = 2u_{n-1} - 8$ where $n \geq 2$

  a. Set nMin to be the n-value of the starting term; in this example enter 1.

  b. Enter the equation for u(n)=. To get $u(n-1)$, press 2nd [u] ((X,T,θ,n − 1)).

  c. Set u(nMin) to be the value of the starting term; in this example enter 47. (The calculator will put the value in braces.)

You can find values of individual terms, as well as a range of terms, on the Home screen. To find u_{22}, press 2nd [u] (22). To find a range of terms, use a comma between the first and last term.

Note 1J • Graphing Sequences

You can graph sequences to display numbers generated by recursive formulas. The x-axis will represent the values of n, and the y-axis will represent the values of $u(n)$.

Go to the Window screen. Set the window values to show the part of the graph you want to see.

  nMin = the smallest value of n you want graphed on the x-axis. You've already set this on the Y= screen.

  nMax = a value a little larger than the greatest value of n you want graphed.

(continued)

PlotStart = the first term of the sequence you want graphed. This is almost always 1.

PlotStep = the terms you want graphed. For example, if you want to plot every other term, PlotStep=2. PlotStep is almost always 1.

Xmin = and Xmax = the minimum and maximum values on the *x*-axis. These usually will be about the same as *n*Min and *n*Max, unless you want a close-up look at some part of the graph.

Xscl = and Yscl = the distance between tick marks on the two axes. The number of divisions should be less than 25. If there are too many tick marks, the axes will appear too thick.

Ymin = and Ymax = the range of function values you want graphed. Usually Ymin will be slightly less than the smallest function value and Ymax will be slightly greater than the largest function value.

Press GRAPH to see the graph.

These screens show graphing 20 terms of each sequence.

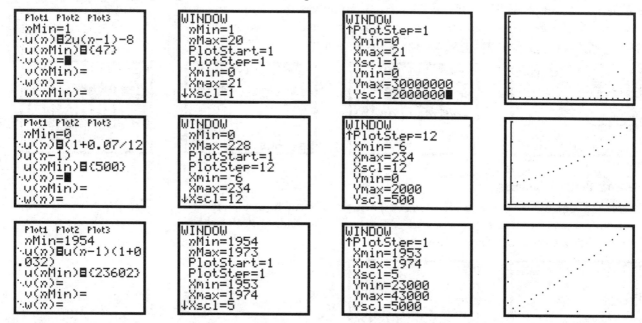

If you enter more than one sequence into the Y= screen, all will be graphed at the same time.

Note 1K • Finding Specific Terms

You can view many elements of a sequence at once by using sequence tables. First enter the sequence into the Y= screen. (See **Note 1I** if you need help entering a sequence.) Then press 2nd [TBLSET]. TblStart is the smallest *n*-value for which you wish to see a sequence value. The value of ΔTbl specifies which terms will actually be displayed. For example, if ΔTbl=3 the table will display every third term. Press 2nd [TABLE] to display the table. Use the up and

(continued)

down arrow keys to see more *x*-values, or the right and left arrow keys to see values of other sequences that are entered.

 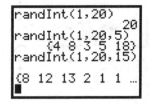 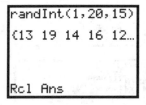

Note 1L • Random Numbers

There are several ways to generate a list of random numbers within an interval.

Random Integers

To find a random integer between 1 and 20, on the Home screen press MATH and arrow to PRB. Select 5:randInt(and enter 1,20), then press ENTER. If you want five random numbers, either press ENTER five times, or enter randInt(1,20,5) and press ENTER. If you ask for more numbers than show on one line of the screen, you can scroll to see the rest of the list. Or you can press 2nd [RCL] 2nd [ANS] ENTER to see the entire list on the screen.

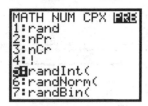 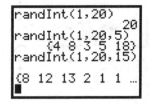 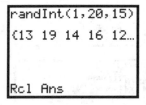 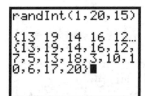

Random Decimal Numbers

Press MATH, arrow to PRB, and select 1:rand. Then press ENTER to display a random decimal number between 0 and 1. To generate a random decimal number between 0 and 8, enter 8*rand.

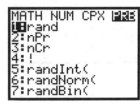 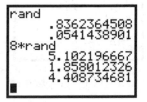

Errors

If you are getting exactly the same random numbers as someone else, try changing the seed value. Enter a number other than 0 and press STO→ MATH, select PRB, and press ENTER.

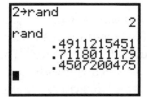

Note 1M • Finance Solver

The Finance TVM (Time Value of Money) solver will solve problems about simple loans, mortgages, and investments. Press APPS and select 1:FINANCE. Choose 1:TVM solver.... Enter values into all but one of the following positions. The solver will then calculate the missing entry. In general, negative amounts indicate money you give to the bank and positive amounts indicate money you receive.

N = the total number of payments.

I% = the annual interest rate as a percent.

PV = the principal or starting value (this is 0 or negative for investments).

PMT = the payment or regular deposit (this is 0 or negative for investments).

FV = the final value.

P/Y = payments per year.

C/Y = interest calculations per year.

PMT:END BEGIN indicates whether payments are made at the end or beginning of each month.

After entering the six known values, highlight the value you want to find and press ALPHA [SOLVE].

This screen shows calculating the monthly payment to completely repay a 5-year (60-month) $12,000 loan at 5.25% interest, with payments made at the end of each month. The answer, PMT, is negative because it is a payment made to the bank.

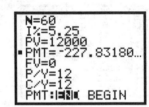

Note 2A • Basic Statistics

You can get several standard statistics for a data set stored in a list. Press [STAT] CALC 1:1-Var Stats, enter the name of the list, and press [ENTER]. If the frequencies of the data values are stored in another list, enter both list names separated by a comma. The following values will be displayed.

$\bar{x}$ = the mean.

Σx = the sum of the x-values.

Σx^2 = the sum of the squares of the x-values.

Sx = the sample standard deviation.

σx = the population standard deviation.

n = the number of data values.

minX = the minimum data value.

Q_1 = the first quartile.

Med = the median.

Q_3 = the third quartile.

maxX = the maximum data value.

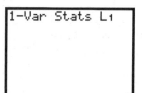 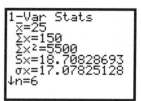 

You can display some of these statistics individually, such as the mean and median. Press [2nd] [LIST] MATH.

Note 2B • Box Plots

The calculator can display up to three box plots at once. First, set up each plot, and then set up the window.

For help setting the Plot Setup screen, see **Note 1F.** There are two types of box plots. The first type marks outliers as special. The calculator uses the standard rule for defining an outlier: Values greater than $Q_3 + 1.5 \cdot IQR$ or less than $Q_1 - 1.5 \cdot IQR$ are outliers. If you select this type of box plot, you must choose what mark to use for an outlier. The second type of box plot does not show outliers as different from other data points.

After setting up all your plots, go to the Window screen to set your graphing window. See **Note 1F** for help determining the window values to use. In a box plot Ymin, Ymax, and Yscl can be any value as long as Ymin is less than Ymax. When you are finished setting your window values, press [GRAPH].

(continued)

Discovering Advanced Algebra Calculator Notes for the Texas Instruments TI-83 Plus and TI-84 Plus

Plot1 below is a box plot using the data set L1 = {0, 10, 20, 30, 40, 50}. Plot2 and Plot3 use the data set L2 = {0, 5, 10, 15, 20, 50}. Data in Plot3 have frequencies L3 = {7, 1, 1, 3, 2, 1}. This plot also shows outliers.

 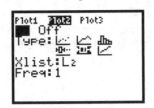  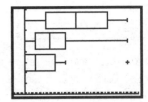

Note 2C • Histograms

To graph a histogram, set the Plot Setup screen as directed in **Note 1F.**

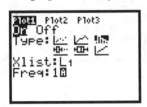

Now set your Window screen. Setting a good window to view a histogram may take several tries. First decide on the bin width to use so that there are not too many or too few bins in your graph. A good rule is to set the bin width at about 15% of the range. Once you have found a good value for the bin width you can make your first try at a window:

Xmin = a multiple of bin width that is less than or equal to your smallest value.

Xmax = a multiple of bin width that is greater than your largest value.

Xscl = bin width.

Ymin = −1. A negative value keeps the tracing values from covering the bins.

Ymax = the number of items in your tallest bin. This will probably be a guess; start with half the number of items in your data set.

Yscl = the distance between tick marks on the *y*-axis. The number you choose will depend on the Ymax value. If the tick marks are too close together the *y*-axis will appear too thick.

Xres = 1. This number does not affect a histogram.

Press [GRAPH]. If the graph doesn't fit well in the window, press [TRACE] and use the left and right arrow keys to find the number of items in the tallest bin. (This is n= in the lower right of the screen.) Go back to the Window screen and reset Ymax (and Ymin if needed) and press [GRAPH] again.

In this example, the first two screens show setting up the stat plot for the data set {22, 25, 25, 27, 27, 27, 28, 28, 29, 31, 32, 32, 37}. The second two screens show how to set up the histogram if the same data are in list L2 and their frequencies (number of occurrences) are in list L3. The latter is usually done only for very large lists.

(continued)

 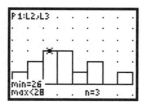  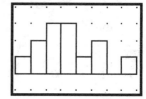

In either case, the range is $37 - 22 = 15$. Fifteen percent of that is $(15)(0.15) = 2.25$, so bin width = 2. There are 13 data points, so Ymax = 6.5. The histogram in the second screen is too short, so the window is adjusted.

Note 2D • Naming Lists

In addition to the six standard lists L1 through L6, you can create more lists as needed. You can also give the standard lists meaningful names (of five or fewer characters) to help you remember what data are where. Here are three ways to name a list.

Naming a List on the Stat Edit Screen

Press [STAT] [ENTER] and arrow up to a list name. Then arrow right to the list that immediately follows the last named list. Its name field will be blank. You are now in Alpha-Lock mode, so type a name for your list and press [ENTER]. You can now enter list values.

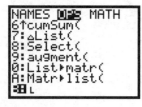 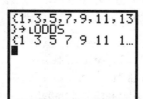 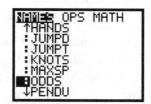 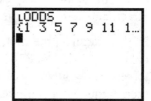

Naming a List on the Home Screen

If you're entering this short list into the Home screen and you want to name it ODDS, type {1,3,5,7,9,11,13} and press [STO→] [2nd] [LIST] OPS, then select B:L. Enter your list name by pressing [ALPHA], the alphabet keys, and then [ENTER]. To refer to that list in the future, press [2nd] [LIST] NAMES, choose the list name, and press [ENTER].

(continued)

Copying and Naming Data from a Standard List

If you have data in a standard list, such as list L1, you can copy the data into another list and then name that list. On the Home screen, press 2nd [L1] STO→ 2nd [LIST] OPS and select B:L. Use Alpha-Lock mode to type a list name, and then press ENTER. To display the list, press 2nd [LIST] NAMES, arrow to your list, and press ENTER ENTER.

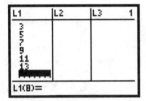  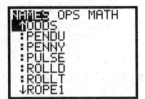 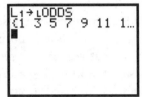

Resetting a List

To reset the calculator so that only lists L1 through L6 are displayed, press STAT 5 (SetUpEditor) ENTER. This action will not delete a named list from the calculator's memory and you will still be able to recall a named list with its stored data.

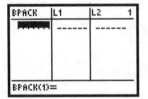 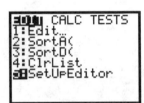 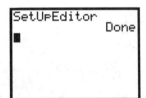 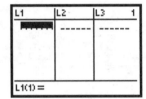

Deleting and Archiving a List

To delete a list, press 2nd [MEM], select 2:Mem Mgnt/Del..., and then 4:List..., arrow to the list you want to delete, and press DEL. You can delete a pre-set list or a named list. If you delete a list, you lose the data in the list. To avoid losing the data, instead of pressing DEL, press ENTER to mark the list with an asterisk. This is called archiving and will temporarily disable the list(s) you mark. An archived list will not appear on the screen when you press STAT 1 (Edit...). By pressing 2nd [LIST] you can see that each archived list is preceded by an asterisk. An archived list retains its data but cannot be used until it is enabled. To enable an archived list, press 2nd [MEM], select 2:Mem Mgnt..., 4:List, arrow to the list you want to enable, and press ENTER. The asterisk disappears.

  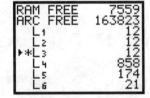 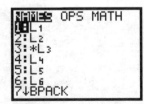

(continued)

Recalling a List

Press $\boxed{\text{STAT}}$ $\boxed{1}$ (Edit...) to display the current lists. Next, create a blank, nameless list by highlighting a list name and pressing $\boxed{\text{2nd}}$ [INS]. To recall one of the pre-set lists (lists L1 through L6) that you previously deleted, press $\boxed{\text{2nd}}$ [L1] or $\boxed{\text{2nd}}$ [L2] and so on, and $\boxed{\text{ENTER}}$. The name will reappear but not the data. (You can recall all the deleted pre-set list names by using the **Resetting a List** procedure.) To recall a previously named list that was hidden from view by resetting a list, press $\boxed{\text{2nd}}$ [LIST], arrow down to the list you want to recall, and press $\boxed{\text{ENTER}}$ $\boxed{\text{ENTER}}$. The list name and data reappear. Note that you cannot recall a list that is archived unless you enable it first. Using the **Resetting a List** procedure will enable lists L1 through L6 whether they are archived or not.

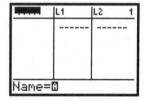 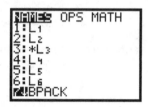

Note 3A • Entering and Graphing Equations

Equations are entered into the Y= screen for various purposes, including graphing. You can enter up to ten equations, Y_1 to Y_0.

Entering Equations

Make sure you are in Function mode. Press MODE. In the fourth line select Func, then press Y=. Enter an equation. The variable must be X using the X,T,θ,n key. You can edit an equation by using the arrow keys and DEL or 2nd [INS]. To remove an equation from the Y= screen, highlight it and press CLEAR.

Before actually graphing, you'll need to determine which part of the Graph screen you want to view.

Setting the Window

Press WINDOW and enter these values.

   Xmin = the minimum *x*-value you want displayed.

   Xmax = the maximum *x*-value you want displayed.

   Ymin = the minimum *y*-value you want displayed.

   Ymax = the maximum *y*-value you want displayed.

   Xscl = and Yscl = the number of units between tick marks on each axis. If there are too many tick marks, individual marks won't be distinguishable and the axes will appear too thick.

   Xres = 1.

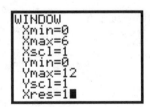

Graphing

Press GRAPH to see the graph of the equation(s). They will appear one after another, in the order listed on the Y= screen.

If the graph is not situated on the screen to your satisfaction, go back to the Window screen and change the values. Experiment with various window settings until you're satisfied with the appearance of the graph.

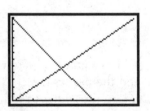

You can turn off the graph of an equation without clearing it from the Y= screen by arrowing to its = symbol and pressing ENTER. When the = symbol is not highlighted, the equation is turned off and will not graph.

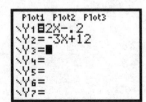
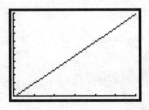

(continued)

Tracing

You can find approximate coordinates of points on the graph by tracing. Press TRACE and a "spider" appears on the first graph. Use the left and right arrow keys to move it along the graph. The coordinates of the spider's position appear at the bottom of the screen. You can move to the graphs of other equations by arrowing up or down.

Notation on the upper-left part of the screen tells you which equation's graph is being traced.

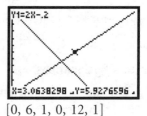

[0, 6, 1, 0, 12, 1]

Zooming

There are several ways to enlarge part of the graph. You can go back to the Window screen and change the window settings, or you can choose one of the commands that appear when you press ZOOM.

1:Zbox allows you to define your own enlargement. Select 1:Zbox to display a spider. Use the arrow keys to move the spider to the area you'd like to enlarge. (This spider isn't restricted to the curve the way the trace spider is.) Press ENTER. Then draw a box by arrowing to the corner diagonally opposite from your current position. Press ENTER again. The interior of the box will enlarge and fill the screen.

2:Zoom In enlarges the screen by a factor of 4. Selecting 2:Zoom In will display a spider that you can position to where you want the center of your new enlarged screen. Press ENTER to see the new screen.

3:Zoom Out does the opposite of Zoom In. Select 3:Zoom Out, position the spider to the desired screen center, and press ENTER to see the new screen.

Zooming automatically changes the settings on the Window screen.

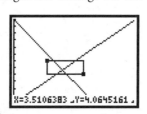

[0, 6, 1, 0, 12, 1]

Graphing a Line and a Plot

You can graph a line over a plot by entering the equation into the Y= screen and the plot as directed in **Note 1F.** If you trace, arrowing up or down causes the spider to jump to each plot and to each function in order.

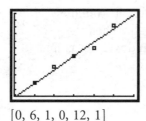

[0, 6, 1, 0, 12, 1]

(continued)

Discovering Advanced Algebra Calculator Notes for the Texas Instruments TI-83 Plus and TI-84 Plus

Setting the Graph Style

In order to distinguish between several displayed graphs or to achieve a special effect, it is sometimes helpful to use a graph style other than the usual thin, solid line.

Use the left arrow key to highlight the style symbol to the left of Y₁= and repeatedly press ENTER to cycle through the various styles. These examples show the possible styles.

   Y₁ graphs a curve using the usual thin, solid line. This is the default setting.

   Y₂ graphs a curve using a thick, solid line.

   Y₃ shades the area above the curve.

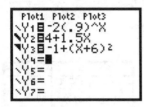 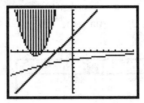

$$[-10, 10, 1, -10, 10, 1]$$

   Y₄ shades the area below the curve.

   Y₅ shows a moving circle that follows the curve and leaves a path.

   Y₆ shows a moving circle that follows the curve but leaves no path (not shown on the screen here).

   Y₇ graphs a curve using a dotted line.

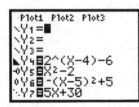 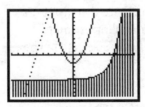

$$[-10, 10, 1, -10, 10, 1]$$

Note 3B • Function Tables

You can build a table of values for any function entered into the Y= screen. Press 2nd [TBLSET].

   TblStart = the first *x*-value you wish to see in the table when first viewed.

   ΔTbl = the difference between the *x*-values in the table. In the first screen here ΔTbl=.1, so the difference between successive *x*-values is 0.1. ΔTbl can be negative.

   Indpnt: set to Auto means that the table will automatically start with the *x*-value equal to the TblStart value. If Indpnt: is set to Ask, the table will be blank until you provide the *x*-values.

   Depend: should always be set to Auto.

(continued)

Press [2nd] [TABLE] to display the table.

 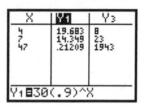  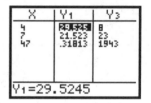

If Indpnt: is set to Auto on the TABLE SETUP screen, you can arrow up or down to see more *x*-values. You can also arrow right to see values of other functions that are turned on in the Y= screen. You can see only two columns of dependent variables at a time.

If you arrow up to the top of a function column, you can see the equation displayed at the bottom of the screen. Press [ENTER] to edit the equation. The changes will be reflected in the table when you press [ENTER] again.

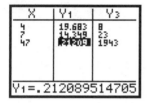

Note 3C/App • Balloon Blastoff Using the EasyData App

You will need a CBR2 (Calculator-Based Ranger).

Connect the CBR2 to the calculator. Press [APPS] and select EasyData. The CBR2 will immediately begin collecting distance data, which are displayed on your calculator screen. However, you need to collect data at shorter intervals for this experiment. Press Setup ([WINDOW]) and make sure 1:Dist is selected. The calculator will ask you to confirm, so press OK ([GRAPH]).

Aim the CBR2 at the rocket and press Start ([ZOOM]). The CBR2 will collect data for about 5 seconds. If you want to stop it sooner, press the trigger. The calculator will display a graph of your data, which you can trace using the left and right arrow keys.

If you need to redo the experiment, press Main ([TRACE]), then press Start ([ZOOM]). You will be told that this function will override the previous data. Press OK ([GRAPH]) and repeat the steps above.

To end the application, press Main ([TRACE]), then press Quit ([GRAPH]). You will get a message telling you where the data are stored. Time data are in L_1, distance data are in L_6, velocity data are in L_7, and acceleration data are in L_8.

Note 3D • Median-Median Line

The calculator can find the equation of the median-median line for a set of data. Press [STAT] CALC 3:Med-Med, then enter the two lists that contain the data, separating them with a comma, and press [ENTER]. The independent variable list should be first. The command's default is to use lists L1 and L2, but it is a good habit to always specify the lists to be used.

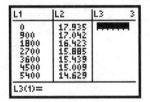  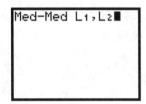 

If you want the equation placed in Y1 on the Y= screen, after the second list press [,] [VARS] Y-VARS 1:Function... 1:Y1 [ENTER].

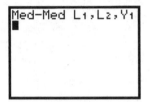 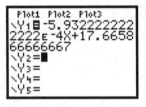 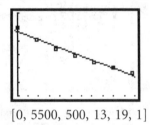

[0, 5500, 500, 13, 19, 1]

Note 3E • Residuals and the Root Mean Square Error

Once you have found a model for paired data, you can calculate the residuals and then the root mean square error.

For this example, assume that your data are stored in lists L1 and L2 and your equation is stored in Y1.

Residuals

a. Press [STAT] [ENTER].

b. Move to the name cell at the top of list L3. Define list L3 as the residuals by entering the expression L2−Y1(L1). To get Y1, press [VARS] Y-VARS 1:Function 1:Y1. The resulting list will not change if you change the data in list L1 or list L2 or the equation in Y1. If you want this list to be dynamic (changing when list L1, list L2, or Y1 changes), enter the expression within quotation marks using [ALPHA] ["].

(continued)

Root Mean Square Error

The root mean square error is defined as

$$s = \sqrt{\frac{\sum_{i=1}^{n}(y_i - \hat{y}_i)^2}{n-2}}$$

The numerator of the fraction is the sum of the squares of the residuals. The denominator is 2 less than the number of elements in list L3.

a. First calculate the residuals in list L3 as described above.

b. Calculate the numerator of the fraction and the value of n. Press STAT CALC 1:1-Var Stats 2nd [L3] ENTER. This puts the sum of the squares of the residuals into a variable called Σx^2 and the number of elements in the residual list into a variable called n.

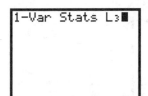 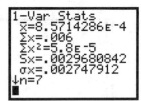

c. Enter this formula into the Home screen: 2nd [√] VARS 5:Statistics:... Σ 2:Σx^2 ÷ ((VARS 5:Statistics 1:n − 2)))) ENTER. The result is the root mean square error.

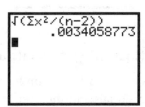

For large values of n, you can find a good approximation by dividing $n - 1$ instead of $n - 2$. This is the sample standard deviation of the residuals, or the value of Sx when you do 1-Var Stats.

Note 3F • Greatest Integer Function

To find the greatest integer less than or equal to a value, press MATH NUM 5:int(, enter the value, and then close the parentheses. If the value is a positive decimal number, the function truncates everything after the decimal point; if the value is a negative decimal number, it does the same and then subtracts 1.

You can also use int(as a function of x. When graphing this function, the calculator may show almost-vertical segments that shouldn't be there. You can eliminate them by changing the graph style to a dotted line. (See **Note 3A** for help setting the graph style.)

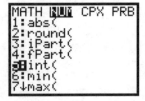 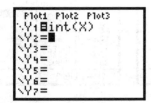 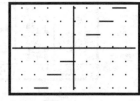

$$[-4.7, 4.7, 1, -3.1, 3.1, 1]$$

Discovering Advanced Algebra Calculator Notes for the Texas Instruments TI-83 Plus and TI-84 Plus

Note 4A • Function Notation

The calculator treats an equation entered into the Y= screen as a function.
A function can be evaluated for different x-values using standard function
notation. For example, $Y_1(5)$ will give the value of the function when x is 5.
On the Home screen press VARS Y-VARS 1:Function… followed by the number
of the equation you want, and the x-value.

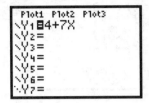   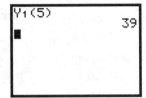

Note 4B • Entering Programs

Your calculator is like a small computer. You can instruct (or program) it
to do calculations or to communicate with other devices. What follows are
not instructions on how to write a program—they are instructions on how
to enter into your calculator a program that has already been written. You
must be very careful to enter the commands exactly as they are written.
Changing the program in any way will alter how it runs or may cause an
error.

There are three ways to enter a program into your calculator. Two easy
ways are to download the program from either a computer or another
calculator. To use a computer, download the program from a CD-ROM or
website directly into your calculator by using TI Connect™ software. You
will need the proper cable to link your computer to your calculator. To
use another calculator that has the program in its memory, link the two
calculators with a link cable and transfer the program as you would a list.
(See **Note 1H**.) The third way is to enter a program manually by following
the steps below:

    a. Press PRGM and arrow to NEW.

    b. Press ENTER and enter the name of the program. You are already in
    Alpha-Lock so do not press ALPHA unless you want to enter a number
    into the program's name.

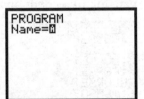

(continued)

c. Enter the program exactly as you see it. Lowercase letters in the commands of the program indicate that you must find its command on the calculator. You can find many programming commands by pressing PRGM and looking under CTL and I/O. To find other commands, functions, and sysbols, either look in the menu of the appropriate calculator key, or press 2nd [CATALOG] and either the letter that begins the command you want. (You are in still in Alpha mode so don't press ALPHA.) Then arrow to the command and press ENTER.

If you enter a command from the calculator keypad using ALPHA characters, the letters will all be uppercase. The command won't look the same as how it's been written and it won't work.

d. Use 2nd [INS] ENTER to insert a new line between two lines. Use DEL on a blank line to delete that line.

e. Press 2nd [QUIT] when you finish entering the program.

Errors

You can edit the program if there is an error or if you need to make a change.

a. Press PRGM and arrow to EDIT. Arrow down to the program and press ENTER.

b. Scroll down through the program to find the error. Studying the entire program on paper is often easier than working from the calculator screen where you can see only a small portion at one time. Use 2nd [INS] or DEL as needed to make changes. Press 2nd [QUIT] when finished.

c. If, when you execute a program, you get an error message that has a Goto option, choose this option. The calculator will automatically switch to EDIT mode and scroll to the line with the error. Then proceed as in **step c.**

Note 4C • Movin' Around

With bits of tape, label two CBR2s A and B. Label two calculators A and B, and connect each to the respective CBR2. Use the CBRSET and CBRGET programs to collect data for 10 seconds.

Make sure you have both programs on both calculators. (See **Note 4B** if you need help entering a program on your calculator.) Run the CBRSET program by pressing PRGM and arrowing to CBRSET. The name of the program will appear on your screen. Press ENTER. The program will prompt you to enter S, the total time to collect data in seconds, and N, the total number of samples, or data points to gather. For this investigation, you should collect data for 10 seconds and have 200 samples.

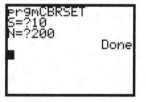

Disconnect the calculators from the CBR2s. When you are ready to collect data, press the trigger on each CBR2. When you have finished collecting data, plug the CBR2s back into the calculators and run the CBRGET program: press PRGM and arrow to CBRGET. On each calculator, the time

(continued)

data will be in list L1 and the distance data will be in list L2. The calculators will display scatter plots.

```
PROGRAM:CBRSET
Prompt S,N
round(S/N,5)→I
If I>0.2:-0.25int(-4I)→I
Send({0})
Send({1,11,2,0,0,0})
Send({3,I,N,1,0,0,0,0,1,1})
```

```
PROGRAM:CBRGET
Send({5,1})
Get(L2)
Get(L1)
Plot1(Scatter,L1,L2,·)
ZoomStat
```

On the Home screen of calculator B, enter L1→L3, press ENTER, enter L2→L4, and press ENTER. This moves calculator B's time and distance data to lists L3 and L4.

Finally, each group member should link to calculator A and copy lists L1 and L2, and link to calculator B and copy lists L3 and L4. (See **Note 1H** for help with linking lists.)

Note 4D • Setting Windows

A *friendly window* scales the *x*-axis to correspond to the Graph screen's width in pixels (94). As a result, when you trace a curve on a friendly window, the spider always falls on points whose *x*-coordinates are "nice" decimal numbers. The *y*-coordinates are computed values and depend on the function being traced; they may or may not be nice decimal values.

If the *y*-axis is scaled so its units are the same as the units on the *x*-axis, then the window will be a "square" window. On a square window there is no distortion of the graph.

One friendly square window whose trace point has *x*-coordinates that are exact tenths can be found by pressing ZOOM 4:Decimal.

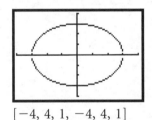
[−4, 4, 1, −4, 4, 1]

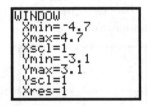
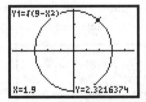

This window is a little small for much of the work in this course. However, if you double the minimum and maximum values in the window screen, you can get a larger friendly square window that is often useful.

You can save the settings for this larger window and recall it at any time. After setting the window values, press ZOOM MEMORY 2:ZoomSto. Now when you want to use it again, press ZOOM MEMORY 3:ZoomRcl. This particular window is often referred to as the friendly window with a factor of 2.

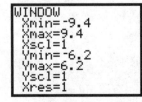
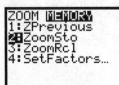
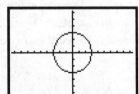

(continued)

It is sometimes helpful to see a grid in the background of the screen display. To turn the grid on (or off), press 2nd [FORMAT] and select GridOn (or GridOff).

   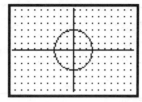

$$[-9.4, 9.4, 1, -6.2, 6.2, 1]$$

Note 4E • Graphing Piecewise Functions

To graph a piecewise function, you add the two functions, with their ranges.

For example, to graph $f(x) = \begin{cases} x + 1 & -5 \leq x < 1 \\ x^2 & 1 \leq x < 3 \end{cases}$, go to the Y= screen and enter Y1=(X+1)(X≥−5)(X<1)+(X^2)(X≥1)(X<3). (To enter the inequality symbols, press 2nd [TEST].) Choose an appropriate window, then press GRAPH.

   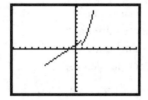

$$[-9.4, 9.4, 1, -6.2, 6.2, 1]$$

Note 4F • Graphing Absolute-Value Functions

To use the absolute-value function, press MATH NUM 1:abs(. For example, to graph $y = |x - 3|$, enter Y1=abs(X−3) into the Y= screen, set an appropriate window, and press GRAPH.

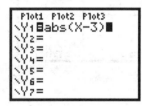   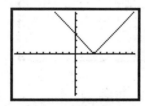

$$[-9.4, 9.4, 1, -6.2, 6.2, 1]$$

Note 4G • Graphing Transformations

The program TRANSFRM gives you practice finding equations for given graphs. From the first menu, choose the type of function you want to practice. In the second menu, you can turn the different transformations on or off by pressing the number key. Press 4 when you're ready. The calculator will display a graph and stop. Study the graph and determine its equation. Press TRACE if you want to see the coordinates of points. When

(continued)

you have decided on an equation, press $\boxed{Y=}$, enter your equation into Y1, and press $\boxed{\text{GRAPH}}$. If your equation is correct, you'll have a match and nothing new will appear on the screen. You can press $\boxed{\text{TRACE}}$ and toggle back and forth between the graph of your function in Y1 and the program's function to confirm that they really do match. If your equation is not correct, the graphs will not match. In that case, press $\boxed{Y=}$ and try again.

When you are finished with one graph, on the Graph screen press $\boxed{\text{CLEAR}}$ $\boxed{\text{ENTER}}$ to run the program again.

         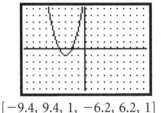

$[-9.4, 9.4, 1, -6.2, 6.2, 1]$

The option 5:GENERIC will draw the graph of a generic function using a thick line and the graph of its image after a transformation using the regular style. Enter the equation of the image into Y1. Use Y8(X) to represent the original function.

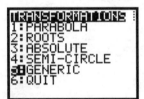   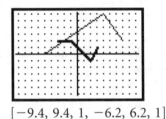   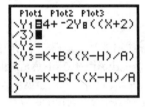

$[-9.4, 9.4, 1, -6.2, 6.2, 1]$

Clean-Up

After you quit the program, you may want to go to the Y= screen and clear the functions so they don't interfere with future work.

```
PROGRAM:TRANSFRM                         If R≠0 and R≠1:0→R
-9.4→Xmin:9.4→Xmax                       If D≠0 and D≠1:0→D
-6.2→Ymin:6.2→Ymax                       Menu("TRANSFORMATIONS","PARABOLA",1,
1→Yscl:1→Xscl                                "ROOTS",2,"ABSOLUTE",3,"SEMI
GridOn:AxesOn                                CIRCLE",4,"GENERIC",5,"QUIT",9)
PlotsOff                                 Lbl 5:F+1→F
"K+B((X-H)/A)²"→Y3                       Lbl 4:F+1→F
"K+B√((X-H)/A)"→Y4                       Lbl 3:F+1→F
"K+Babs((X-H)/A"→Y5                      Lbl 2:F+1→F
"K+B√((1-((X-H)/A)²)"→Y6                 Lbl 1:1→A:1→B
"K+BY8((X-H)/A)"→Y7                      ClrHome:0→G:0→H:0→K
"2(X<-1)/(X≥-3)+(1-X)(X≥-1)              Disp sub("PARABOLA     SQUARE
   (X<2)+(-5+2X)(X≥2)/(X≤3)→Y8              ROOTSABSOLUTE  VALSEMICIRCLE
GraphStyle(8,2)                            GENERIC       ",12F-35,12)
Lbl 0:3→F:                               Repeat G=82
If T≠0 and T≠1:1→T                       Output(3,1,"1.TRANLATE   :"+sub
```

(continued)

(PROGRAM: TRANSFRM continued)

```
  ("OFFON ",3T+1,3))              End
Output(4,1,"2.REFLECT    :"+sub   If R:Then
  ("OFFON ",3R+1,3))              If rand <0.5:-1→A
Output(5,1,"3.STRETCH    :"+sub   If rand <0.5:-1→B
  ("OFFON ",3D+1,3))              End
Output(6,1,"4. GO")              If D:Then
getKey→G:                         A*randInt(1,5)→A
If G=92:1-T→T                     B*randInt(1,4)→B
If G=93:1-R→R                     End
If G=94:1-D→D                     FnOff
End                               FnOn F
If T:Then                         If F=7:FnOn 8
randInt(-7,7)→H                   DispGraph
randInt(-4,4)→K                   Lbl 9
```

Note 4H • Transformations and Compositions

You can use functions entered into the Y= screen in other functions to show transformations and to construct compositions.

Transformations of Functions

You can enter an equation into the Y= screen and then define a second equation as a transformation of the first. For example, enter $4-X^2$ into Y_1 and define Y_2 as $Y_2=3*Y_1(X-4)+2$. (To get Y_1, press VARS Y-Vars 1:Function 1:Y_1.) Y_2 is the image of Y_1 after being stretched vertically by a factor of 3, translated right 4 units and up 2 units.

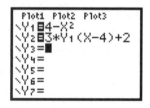 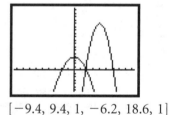

[−9.4, 9.4, 1, −6.2, 18.6, 1]

Compositions of Functions

If you enter two (or more) equations into the Y= screen, you can define another equation as the composition of the equations you have entered. For example, enter $4-X^2$ into Y_1 and $X+5$ into Y_3. Define Y_4 as the composition of Y_1 and Y_3 by entering $Y_4=Y_1(Y_3(X))$.

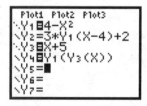 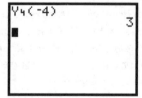 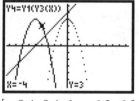

[−9.4, 9.4, 1, −6.2, 6.2, 1]

(continued)

You can use the Home screen recursive loop Y₁(X)→X (or simply Y₁→X) to evaluate the repeated composition of a function with itself. Store a starting value in X and then press [VARS] Y-VARS 1:Function 1:Y₁ [STO→] [X,T,θ,n] [ENTER] [ENTER] [ENTER]. . . . (See **Note 1B** for more on Home screen recursion.)

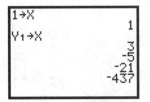

Note 4I • Drawing Segments

On the Graph screen you can draw an overlay on top of the graph.

Follow these steps to draw a segment:

    a. Press [2nd] [DRAW] 2:Line(.

    b. Arrow to one endpoint of the segment you want and press [ENTER].

    c. Arrow to the other endpoint and press [ENTER] again.

Pressing [ENTER] twice ends one segment and begins another at the same point, so you can make a closed figure.

 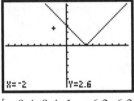 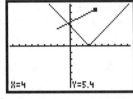 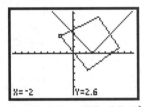

$[-9.4, 9.4, 1, -6.2, 6.2, 1]$ $[-9.4, 9.4, 1, -6.2, 6.2, 1]$ $[-9.4, 9.4, 1, -6.2, 6.2, 1]$

You can also draw segments by entering instructions into the Home screen. To draw a segment between (1, 3.64) and (7.4, 3.64), enter Line(1,3.64,7.4,3.64).

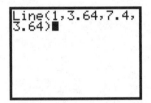 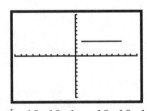

$[-10, 10, 1, -10, 10, 1]$

To erase all drawings, press [2nd] [DRAW] 1:ClrDraw.

Note 4J • Web Graphs

Follow these steps to display a web graph:

    a. Set your calculator to Sequence mode.

    b. Press [2nd] [FORMAT] and select Web in the first line.

    c. Enter the function into the Y= screen. Replace x with u($n-1$) and set u(nMin) to the starting value of x.

    d. Set the Window screen and press [GRAPH].

    e. Press [TRACE]. Each time you press the right arrow key the graph will make one of the two steps in the next iteration of the function.

(continued)

You can clear the web by pressing [2nd] [DRAW] 1:ClrDraw.

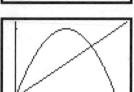 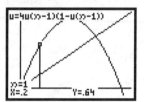 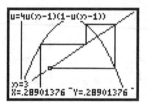 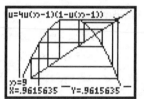

Note 5A • Powers and Roots

Powers

The calculator has special keys for squaring, x^2, and raising to a power of -1, x^{-1} (taking the reciprocal). The command for cubing is found under MATH 3:³. You can find all powers, including these, by using the "caret" key, ∧. If you raise a number to a fractional power, use parentheses around the exponent.

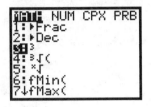 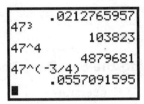

Roots

You can find the square root of a number by pressing 2nd [√]. To find other roots, press MATH. The cube root is 4:$\sqrt[3]{\ }$(and the xth root is 5:$\sqrt[x]{\ }$. For example, to find the fourth root of 47, press 4 MATH 5:$\sqrt[x]{\ }$ 4 7 ENTER.

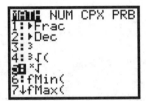 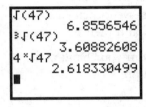

Note 5B • Drawing the Inverse of a Function

Your calculator can draw the inverse of any function. From the Home screen, press 2nd [DRAW] 8:DrawInv, followed by an expression containing X or one of the functions Y₁ through Y₀. Then press ENTER (not GRAPH).

A drawing is overlaid on top of the Graph screen; it cannot be traced. If you change the window or alter anything on the Y= screen, the drawing will be cleared. You can restore it by returning to the Home screen and pressing ENTER again. To clear a drawing, press 2nd [DRAW] 1:ClrDraw.

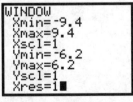 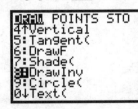  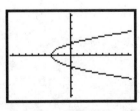

 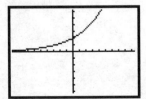 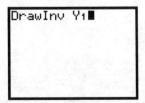 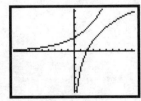

Note 5C • Logarithms and Antilogs

Use $\boxed{\text{LOG}}$ to find the common, or base 10, logarithm of any positive value.
Use $\boxed{\text{2nd}}$ [10ˣ] for the common antilog of a number. Pressing $\boxed{1}$ $\boxed{0}$ $\boxed{\wedge}$ $\boxed{(}$
gives the same result as pressing $\boxed{\text{2nd}}$ [10ˣ].

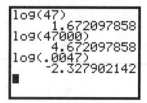
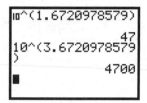

Note 5D/App • Gathering Temperature Data Using the EasyData App

You will need a CBL2 (Calculator-Based Laboratory) and a temperature probe.

Connect the CBL2 and the calculator. Plug the temperature probe into Channel 1. Press $\boxed{\text{APPS}}$ on the calculator and choose EasyData. After some time, the app will recognize the temperature probe.

Press Setup ($\boxed{\text{WINDOW}}$), choose Time Graph, and press $\boxed{\text{ENTER}}$. Press Edit ($\boxed{\text{ZOOM}}$). Enter 10 for the time between samples and press Next, then enter 18 for the total number of samples and press Next. Press OK and then press Start.

Remove the temperature probe from the hot water or from your hand. After the data collection is finished, press Main, press Quit, and press OK.

Note 6A • Entering and Editing Matrices

Entering a Matrix

To enter a matrix, follow these steps:

    a. Press 2nd [MATRX] and from the EDIT submenu select a matrix.

    b. Enter the dimensions of the matrix (rows and then columns).

    c. Enter a value into each cell. Press ENTER to register each entry and to move the cursor to the next position. You can use fractions and operations when you enter values.

    d. When you finish entering values, press 2nd [QUIT] to return to the Home screen.

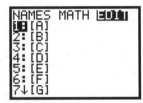  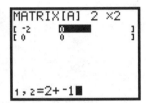 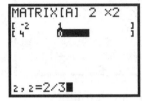

Editing a Matrix

To edit a matrix, follow these steps:

    a. Press 2nd [MATRX] and from the EDIT submenu select the matrix you want to edit.

    b. Arrow to the cell you want to change. Enter the new value and press ENTER. You can also change the dimensions of a matrix. Notice that when you create a new row or column the values in the cells begin as zeros.

    c. When you finish editing values, press 2nd [QUIT] to return to the Home screen.

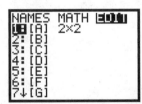 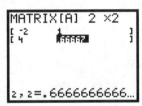 

Viewing a Matrix on the Home Screen

To view a matrix on the Home screen, press 2nd [MATRX] and from the NAMES submenu select the name of the matrix. Press ENTER to display the matrix. If the matrix is too large to fit on the screen, use the arrow keys to scroll across or down the matrix.

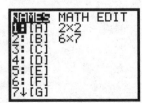 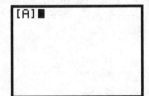 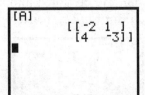 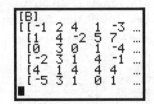

Note 6B • Matrix Operations

You can perform operations with matrices just as with numbers. The following examples use matrices [A], [B], and [C].

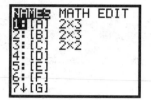 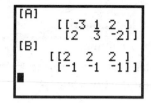 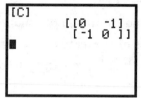

You can add or subtract matrices if they have the same dimensions.

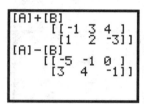

You can multiply two matrices if the number of columns in the first matrix matches the number of rows in the second matrix.

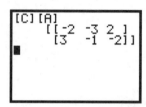

You can multiply any matrix by a constant.

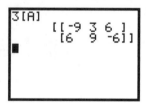

You can raise a square matrix to a power.

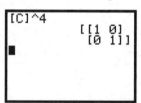

The result of a matrix operation can be stored into a matrix or used in the next calculation. This way you can work recursively with matrices.

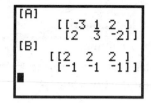

(continued)

Discovering Advanced Algebra Calculator Notes for the Texas Instruments TI-83 Plus and TI-84 Plus

Errors

If you get an ERR:DIM MISMATCH message, then the dimensions of the matrices do not satisfy the operation's criteria.

An ERR:UNDEFINED message probably indicates that you have named a matrix that is not defined.

Note 6C • Plotting a Polygon

You cannot plot a polygon directly from a matrix, but you can convert a matrix into lists and plot a polygon from the lists.

For example, the matrix $\begin{bmatrix} 1 & -2 & -3 & 2 & 1 \\ 2 & 1 & -1 & -2 & 2 \end{bmatrix}$ represents the quadrilateral with vertices $(1, 2)$, $(-2, 1)$, $(-3, -1)$, and $(2, -2)$.

(To graph a closed figure, the first point must be repeated as the last point.) You can convert the matrix columns into lists by selecting 2nd [MATRX] MATH 8:Matr>list(. However, to plot a polygon, the matrix rows need to be converted into lists of x- and y-coordinates. To switch the rows for columns of your matrix, press 2nd [MATRX] MATH 2:ᵀ. This matrix is the *transpose* of the original matrix.

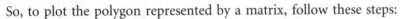

So, to plot the polygon represented by a matrix, follow these steps:

a. Enter the matrix and store it as matrix [A].

b. To store the coordinates as lists, enter 2nd [MATRX] MATH 8:Matr>list(2nd [MATRX] NAMES 1:[A] 2nd [MATRX] MATH 2:ᵀ , 2nd [L1] , 2nd [L2]) ENTER.

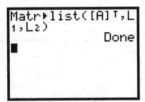

c. Set up Plot1 as an xyline plot with list L1 as the Xlist and list L2 as the Ylist.

d. Set an appropriate window and display the graph.

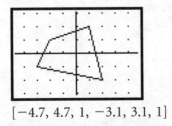

$[-4.7, 4.7, 1, -3.1, 3.1, 1]$

(continued)

You can also use matrices to transform polygons.

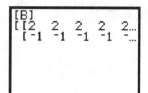 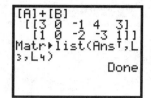 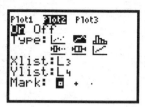 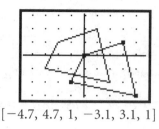

$$[-4.7, 4.7, 1, -3.1, 3.1, 1]$$

Note 6D • Inverse Matrices

To find the inverse of a matrix, enter the name of the matrix and press $\boxed{x^{-1}}$.

If you get an ERR:INVALID DIM message, the matrix is not square; if you get an ERR:SINGULAR MAT message, one row of the matrix is a multiple of another row. In either case, the matrix has no inverse.

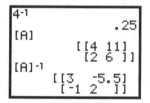

Note 6E • Matrix Row Operations

The calculator can perform four operations on the rows of a matrix.

To exchange two rows in one matrix, use $\boxed{\text{2nd}}$ [MATRX] MATH C:rowSwap. For example, you exchange rows 1 and 2 of matrix [A] with the command rowSwap([A],1,2).

To add the entries of one row to those of another row, use $\boxed{\text{2nd}}$ [MATRX] MATHD:row+(. For example, you add the entries of row 1 to those of row 2 and store them into row 2 with the command row+([A],1,2).

To multiply the entries of one row by a value, use $\boxed{\text{2nd}}$ [MATRX] MATH E:*row(. For example, you multiply the entries of row 1 by 5 and store them into row 1 with the command *row(5,[A],1).

To multiply the entries of one row by a value and add the products to another row, use $\boxed{\text{2nd}}$ [MATRX] MATH F:*row+(. For example, you multiply the entries of row 1 by 5, add the products to row 2, and store them into row 2 with the command *row+(5,[A],1,2).

These commands don't change matrix [A]; they create a new matrix. You'll probably want to end each command by storing the new matrix with a new name or by replacing [A] with the new matrix, as was done in each of the examples.

Note 6F • Reduced Row-Echelon Form

To convert an augmented matrix to reduced row-echelon form, enter
[2nd] [MATRX] MATH B:rref(and the name of the matrix.

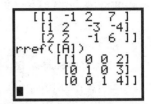

This example shows solving the system

$$\begin{cases} x - y + 2x = 7 \\ x + 2y - 3z = -4 \\ 2x + 2y - z = 6 \end{cases}$$

to get $x = 2$, $y = 3$, and $z = 4$.

Note 6G/App • Graphing Inequalities with the Inequal App

To start the application, press [APPS] and select Inequal. Go to the Y= screen.

    a. Move the cursor over the = symbol.

    b. Press [ALPHA] and one of the five top-row keys, [F1] to [F5], to select the type of inequality you want to graph.

    c. Arrow to the right of the inequality symbol and enter the rest of the inequality.

    d. Set an appropriate window. For ShadeRes= enter an integer from 3 to 8 to adjust the space between the shading lines. The larger the number, the larger the space.

    e. Press [GRAPH] to display the graph of the inequality. Notice that the boundary line of a strict inequality, < and >, is represented with a dashed line.

  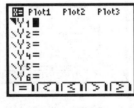 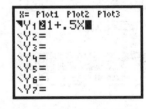

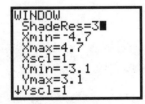 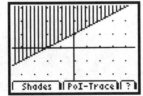 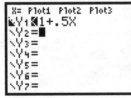 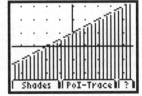

To graph an inequality with a vertical boundary line, arrow to X= in the upper-left corner of the Y= screen, press [ENTER], and proceed as if on the Y= screen.

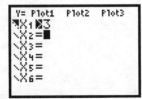 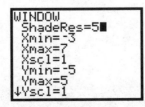 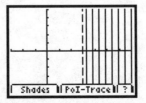

To turn off the application, press [APPS], select Inequal, and choose 2:Quit Inequal.

(continued)

Graphing Systems of Inequalities

a. Enter the system of inequalities, set up the window, and press GRAPH.

b. To find the intersection of the regions, press ALPHA and one of the keys under Shades, [F1] or [F2]. Then select 1:Ineq Intersection.

c. To find the points of the intersection of the boundary lines, press ALPHA and one of the keys under PoI-Trace: [F3] or [F4]. Use the left and right arrow keys to move to a point on the same line and the up and down arrow keys to move to a point on a different line.

These screens show how to graph the region defined by the system

$$\begin{cases} y \leq 6 - x \\ y \leq 5 \\ x \leq 4 \\ y \geq 0 \\ x \geq 0 \end{cases}$$

The last screen shows the intersection of the two boundary lines Y_1 and Y_2.

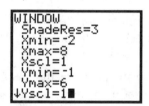
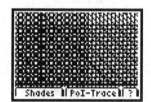

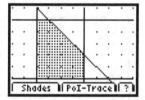
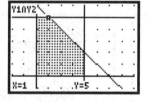

Discovering Advanced Algebra Calculator Notes for the Texas Instruments TI-83 Plus and TI-84 Plus

Note 7A • Free Fall

Use the CBRSET and CBRGET programs to collect data for this investigation. (See **Note 4C** if you need help with these programs.) In CBRSET, use a value between 2 and 5 for S and make sure you collect 20 samples/second (set N to a value between 40 and 100). Then disconnect the calculator.

Position the CBR facing up on the floor as described in the investigation instructions. Press the trigger on the front of the CBR to collect data.

When data collection is complete, press ENTER and view the data to see if there is a short section of data showing the drop. If not, make the needed modifications to your procedure and repeat the data collection.

Use TRACE to identify the points of the drop, and copy these *x*- and *y*-values onto your paper. Continue with the investigation.

Note 7B • Finite Differences

To calculate finite differences for a list, enter 2nd [LIST] OPS 7:ΔList(and the name of the list. You can use this command recursively to look at the values of the first, second, and third differences of a sequence.

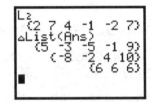

Data Analysis Using Finite Differences

You can also use the command ΔList(on the Stat Edit screen to create lists of differences that can then be graphed.

    a. Enter *x*-values into list L1 and *y*-values into list L2. Remember that the *x*-values must be an arithmetic sequence.

    b. Set a window (or use ZoomStat) and display a scatter plot of (L1, L2).

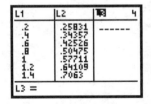   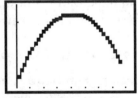

The plot does not show a horizontal linear pattern or even a linear pattern, so proceed to look at a graph of the first differences.

    c. Define list L4 to be the first difference of list L2, ΔList(L2). Note that there is one less element in list L4 than there is in list L2.

    d. Let list L3 be the same as list L1 but omit the first entry. Lists L3 and L4 must have the same number of elements.

    e. Set a window (or use ZoomStat) and display a scatter plot of (L3, L4).

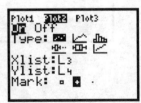  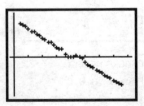

(continued)

The plot shows a linear but nonhorizontal pattern using the first differences, so you must proceed to the next set of differences.

  f. Define list L6 to be the differences of list L4, which are the second differences of the original y-values in list L2. There is again one less element in list L6 than there is in list L4.

  g. Let list L5 be the same as list L3 but, again, omit the first element.

  h. Set a window (or use ZoomStat) and display a scatter plot of (L5, L6).

 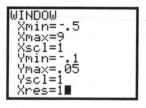 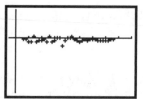

This graph of second differences has a horizontal linear trend, so the data in lists L1 and L2 can be modeled with a 2nd-degree polynomial.

Note 7C • Rolling Along

Use the CBRSET and CBRGET programs to collect 6 seconds of data. (See **Note 4C** for help with these programs.)

Position the CBR at the low end of the table as described in the investigation instructions.

When data collection is complete, view the data to see if they look like a parabola. If not, make the needed modifications to your procedure and repeat the data collection.

Continue with the investigation. As mentioned in the instructions, make sure to subtract 0.5 meter from the distance measurements in list L2.

Note 7D • QUAD Program

The program QUAD requests values A, B, and C. These values refer to the coefficients of the quadratic equation $ax^2 + bx + c = 0$. The solutions to the equation are displayed on the screen and stored in R and S.

If your calculator is in Real mode and your equation has no real roots, then the program will give an error message. (See **Note 7E** for alternatives to Real mode.)

 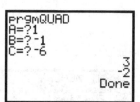 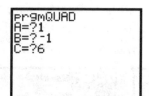 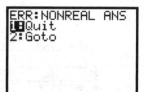

```
PROGRAM:QUAD
Prompt A,B,C
(-B+√(B² - 4AC))/(2A)→R
(-B-√(B² - 4AC))/(2A)→S
Disp R,S
```

Note 7E • Complex Numbers

Your calculator has three number display modes. Press MODE and look at the seventh line.

Real   Displays only real values unless a complex number using *i* is entered. Otherwise, it gives an ERR:NONREAL ANS message.

a+bi   Displays both real and nonreal values in the form $a + bi$. You will use this mode when working with complex numbers.

re^θi   Displays both real and complex values in polar form. (Not used in this course.)

When entering a complex number, you get the symbol *i* by pressing 2nd [*i*], or if you are in a+bi mode, by entering $\sqrt{(-1)}$.

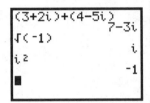
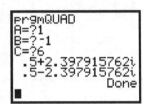

Note 7F • MANDELBR Program

The program MANDELBR is based on the principle that, if a point moves more than 2 units from the origin, it will never return. If it remains within 2 units of the origin after 50 iterations, then it will likely stay in that range, so the program turns on the corresponding pixel.

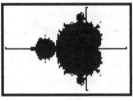

$[-2, 1, 1, -1, 1, 1]$

First set the window you want to explore. To view the entire Mandelbrot set, set the window to $[-2, 1, 1, -1, 1, 1]$. Then run the program MANDELBR. Running this program can take several hours.

```
PROGRAM:MANDELBR                          0→N:0→Z
a+bi                                      While N≤50 and abs(Z)≤2
PlotsOff                                  N+1→N
ClrDraw                                   Z²+A+Bi→Z
FnOff                                     End
RectGC:CoordOn:GridOff:AxesOn             If abs(Z)≤2
DispGraph                                 Pt-On(A,B)
For(A,Xmin,Xmax,(Xmax+Xmin)/94)           End
For(B,Ymin,Ymax,(Ymax-Ymin)/62)           End
```

Note 7G • SYNDIV Program

The program SYNDIV performs the synthetic division $\frac{P(x)}{x-a}$.

    a. Run the program SYNDIV.

    b. Enter the coefficients of the divisor, polynomial $P(x)$, as a list. These values must be entered in order of descending degree, including zeros for missing terms, and they must be separated by commas and enclosed in braces, { }. Press ENTER.

    c. For VALUE, enter the value of a in the divisor $x - a$ as an integer, decimal, or fraction. Press ENTER.

    d. The program displays the coefficients of $P(x)$ that you entered, the middle row of the division, and then the coefficients of the quotient polynomial in the third line. It displays the remainder of the division in the last line.

 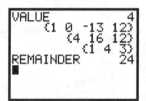

After you have run the program, you can press ENTER to run it again. If you want to use the same polynomial, $P(x)$, again, enter list L1 for the coefficients. If you want to use the result of the last division as the new dividend, then enter list L3 for the coefficients of your new polynomial.

```
PROGRAM:SYNDIV                          XR→L2(J)
Disp "ENTER COEFFIC.","USE L1 FOR       R→L3(J)
   SAME","OR L3 FOR RESULT"             End
Input L1                                L1(J)+XR→R
Input "VALUE:",X                        ClrHome
dim(L1)-1→D                             Disp X,L1,L2,L3,R
0→R:{0}→L2: L2→L3                       Output(1,1,"VALUE")
For(J,1,D)                              Output(5,1,"REMAINDER")
L1(J)+XR→R
```

Note 7H • Zero Finding

Your calculator can find the zero(s) of a function.

    a. Enter the equation into the Y= screen.

    b. Find a window that shows the zero you want to determine and display the graph.

    c. Select 2nd [CALC] 2:zero.

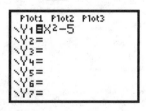 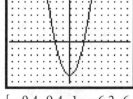 

$$[-9.4, 9.4, 1, -6.2, 6.2, 1]$$

(continued)

d. The calculator prompts you to enter left and right bounds around the
zero and a guess. You can do this by arrowing left or right and pressing
ENTER, or by typing a number.

Left Bound = an *x*-value that is to the left of the zero.

Right Bound = an *x*-value that is to the right of the zero.

Guess = an *x*-value that is very near the zero.

e. The calculator locates a zero between the left and right bounds, if
one exists.

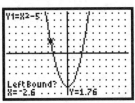 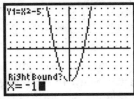 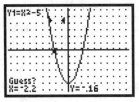 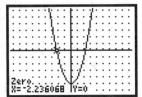

$[-9.4, 9.4, 1, -6.2, 6.2, 1]$

Note 8A • Intersections, Maximums, and Minimums

Intersections

Follow these steps to find an intersection of two curves without tracing:

a. Display the graph of both curves.

b. Press [2nd] [CALC] 5:intersect.

c. The screen shows the two curves with the spider on the curve defined by Y_1. The prompt calls for a First Curve?. If you have graphed more than two curves and Y_1 does not define one of the curves you want, use the up and down arrow keys to select a different curve. Press [ENTER].

d. The prompt then calls for a Second Curve?. If necessary, use the up and down arrow keys to select a curve, and then press [ENTER].

e. Finally, the prompt calls for a Guess? point. Use the left and right arrow keys to move the spider near the intersection you want to find, and then press [ENTER]. Note: If the two curves have more than one intersection, you must confine yourself to the vicinity of the intersection you want.

f. The screen shows the coordinates of the intersection nearest your guess.

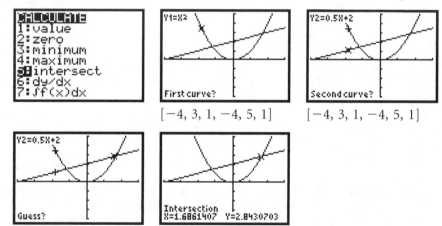

$[-4, 3, 1, -4, 5, 1]$ $[-4, 3, 1, -4, 5, 1]$

$[-4, 3, 1, -4, 5, 1]$ $[-4, 3, 1, -4, 5, 1]$

Maximums and Minimums

A similar process allows you to find the coordinates of a maximum or minimum without tracing. For example, follow these steps to find the minimum of $y = (x - 3)^2 + 4$:

a. Display the graph of the function.

b. Press [2nd] [CALC] 3:minimum.

c. The prompt calls for a Left Bound?. Move the spider to the left of the minimum and press [ENTER]. Note: If the curve has several extreme values, you must confine yourself to the vicinity of the maximum or minimum you want.

d. The prompt calls for a Right Bound?. Move the spider to the right of the minimum and press [ENTER].

e. Finally, the prompt asks for a Guess?. Move the spider between the two bounds and press [ENTER].

(continued)

f. The screen shows the coordinates of the minimum between the specified bounds.

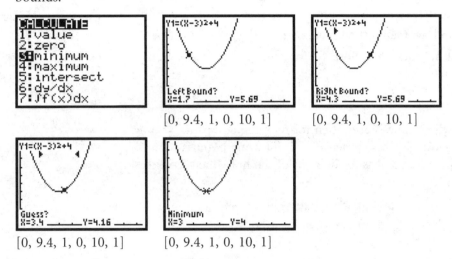

[0, 9.4, 1, 0, 10, 1]   [0, 9.4, 1, 0, 10, 1]

[0, 9.4, 1, 0, 10, 1]   [0, 9.4, 1, 0, 10, 1]

Selecting 4:maximum results in the maximum between the specified bounds.

Note 8B • Asymptotes, Holes, and Drag Lines

When the calculator graphs a function, it plots a sequence of points and connects each point to its adjacent points. The first point's x-coordinate equals Xmin. Each subsequent point's x-coordinate increases by $\frac{1}{94}$ of the screen width. If you graph a function that has a vertical asymptote or a hole that falls between two consecutive plot points, the calculator may not properly display the graph.

Asymptotes

If the graph has a vertical asymptote that falls between two consecutive plot points, the calculator sometimes draws an erroneous, almost vertical, drag line. This occurs because the calculator connects the two points that span the asymptote, one with a positive y-coordinate and the other with a negative y-coordinate.

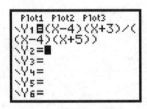 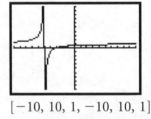

[−10, 10, 1, −10, 10, 1]

To eliminate the drag line, you must use a window in which the x-coordinate of one of the plot points equals the asymptote value. Most vertical asymptotes that you will encounter have simple, rational values. Therefore, an appropriate friendly window, which plots points with "nice," rational coordinates, will often attempt to plot a point on the asymptote. Such a point is undefined and the calculator cannot connect it to its adjacent points, so there is no drag line.

(continued)

While a friendly window with a factor of 2 will work for many functions with vertical asymptotes, you may have to try other friendly windows from time to time. Remember that Xmin and Xmax are the important values when determining a friendly window; the plotting increment, $\frac{\text{Xmax} - \text{Xmin}}{94}$, must be equal to a "nice," rational number. (See **Note 4D** for more information about friendly windows.)

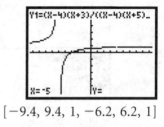

$[-9.4, 9.4, 1, -6.2, 6.2, 1]$

Holes

The graph of a function with a hole is also often misrepresented on the calculator. If, as the calculator plots its sequence of points, one point is to the left of the hole and the next point is to the right of the hole, the segment connecting these points will cover the hole.

To see the hole, you must choose a window that attempts to plot a point in the hole. This usually requires an appropriate friendly window. In the last screen above, you can see the hole at $x = 4$.

Note 9A • Partial Sums of Series

Follow these steps to find partial sums of any recursively defined sequence:

    a. Press MODE and set the fourth line to Seq and the fifth line to Dot.

    b. Press 2nd [FORMAT] and set the first line to Time.

    c. On the Y= screen, enter

nMin = 1.

u(n) = the recursive rule for the sequence.

u(nMin) = the starting value of the sequence.

v(n) = the sum of $v(n-1)$ and the recursive rule for $u(n)$. Sequence v
        is the sequence of partial sums of the terms of sequence u.

v(nMin) = the same starting value as u(nMin).

    d. Set an appropriate Window screen or Table Set screen in order to view the
terms of sequence u and the partial sums, sequence v.

See **Notes 1I** and **1J** for more help with entering or graphing recursive
sequences.

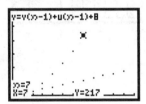 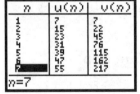

[0, 12, 1, 0, 300, 50]

Note 10A • Dice Simulation

(If your calculator has the application Prob Sim, see **Note 10A/App** for an alternative way to simulate dice.)

Recall that you can simulate the throw of a die using the random integer command, randInt(1,6,n), where n is the number of throws. See **Note 1L** for help with the randInt(command. To store the outcomes into a list, say list L1, press [STO→] [2nd] [L1].

Follow these steps to simulate the sums for 300 throws of a pair of dice:

a. Store 300 throws of a die into list L1, randInt(1,6,300)→L1.

b. Store 300 throws of a die into list L2, randInt(1,6,300)→L2.

c. Define list L3 as the sum of lists L1 and L2.

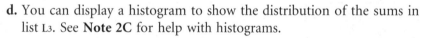

Notice in the second screen that the definition of list L3 uses quotation marks, [ALPHA] ["], and the list name has a diamond, ♦, beside it. The quotation marks make the definition dynamic so that the values in list L3 will automatically update if list L1 or list L2 changes. The diamond indicates that the list is dynamic.

d. You can display a histogram to show the distribution of the sums in list L3. See **Note 2C** for help with histograms.

[2, 13, 1, 0, 70, 10]

Note 10A/App • Dice Simulation with the Prob Sim App

(See **Note 10A** if your calculator does not have the application Prob Sim.)

To start the application, press [APPS], select Prob Sim, and press any key. Follow these steps to simulate the sums for 300 throws of a pair of dice:

a. From the Simulation menu, select 2.Roll Dice.

b. Press [ZOOM] to go to the Settings menu. Enter these settings and then press [GRAPH] to choose OK:

Trial Set:300   The number of trials to perform at once.

Dice:2   The number of dice to use.

Sides:6   The number of sides on each die.

Graph:Freq   The graph can show frequency or probability.

StoTbl:All   The table can store all, the last 50, or none of the trials.

ClearTbl:Yes   The data clear when you do the experiment again.

Update:50   The number of trials after which the graph updates.

c. Press [WINDOW] to roll the dice. The application will simulate 300 throws of a pair of dice and will show a bar graph of the sums. The bar graph will update every 50 rolls.

(continued)

  d. When the 300 throws are complete, you can arrow left or right to trace the bar graph and see the frequency of each sum.

  e. If you press GRAPH, the bar graph will change to a table. You can arrow up or down to see the number on each die, D1 and D2, as well as the sum. Pressing GRAPH again changes the table back to a bar graph.

  f. If you press TRACE, you have the option to save the data into four lists: ROLL for the roll number, D1 for the numbers on die 1, D2 for the numbers on die 2, and SUM for the sum of the dice. Press GRAPH to save the data, or press Y= to escape without saving.

  g. Exit the program by pressing Y= to escape the dice simulation. Press Y= again to remove the trials from memory, and then press GRAPH to quit and Y= to confirm.

As is obvious from the Simulation menu, you can use the Prob Sim application to simulate many other probability situations. When you are in the Settings menu, press WINDOW to set advanced settings, such as the "weight" of a side, which can make the probability of one event greater than another.

  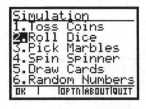 

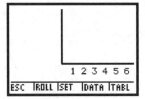 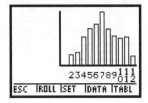 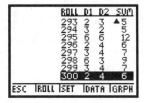

Note 10B • Permutations

To find numbers of permutations, use the nPr command. To find the nPr command, press MATH PRB 2:nPr. First enter the value of *n*, the number of objects. Then enter the nPr command, and enter the value of *r*, the number of objects chosen. Then press ENTER.

For example, to find the number of arrangements of 5 objects chosen 3 at a time, enter 5 nPr 3. The answer shows that there are 60 arrangements.

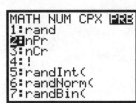 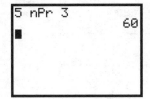

Note 10C • Factorials

To find the factorial command, press MATH PRB 4:!. For example, to find 5!, press 5 MATH PRB 4:! ENTER.

In the order of operations, factorial has higher precedence than negation, so $-3!$ is equivalent to $-(3!)$.

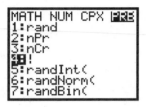 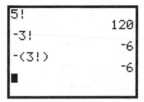

Note 10D • Combinations

To find numbers of combinations, use the nCr command. To find the nCr command, press MATH PRB 3:nCr. First enter the value of *n*, the number of objects. Then enter the nCr command, and enter the value of *r*, the number of objects chosen. Then press ENTER.

For example, to find the number of groupings of 5 objects chosen 3 at a time, enter 5 nCr 3. The answer shows that there are 10 different groupings.

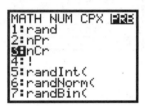 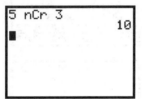

Note 10E • Binomial Probability

Single Probability

To calculate the probability of any number of successes in a probability experiment, use the binomial probability distribution function command, binompdf(. To find the binompdf(command, press 2nd [DISTR] A:binompdf(.

The binompdf(command requires three arguments: the number of trials, the probability of a success for each trial, and the number of successes.

For example, binompdf(10,.75,8) finds the probability of 8 successes out of 10 trials where the probability of each success is 0.75.

The binompdf(command is a shortcut for calculating the value of one term of a binomial expansion. That is, binompdf(10,.75,8) is the same as $_{10}C_8 \cdot (0.75)^8 \cdot (0.25)^2$.

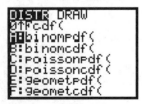 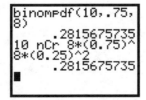

(continued)

To find more than one probability at the same time, use the binompdf(command and enter the number of successes as a list.

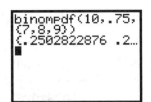

Cumulative Probability

The binomial cumulative distribution function command, binomcdf(, is similar to the binompdf(command, but it sums the binomial probabilities from 0 successes to the desired number. To find the binomcdf(command, press [2nd] [DISTR] B:binomcdf(.

For example, binomcdf(10,.75,6) finds the probability of 6 or fewer successes out of 10 trials where the probability of each success is 0.75. To find the probability of more than 6 successes, subtract the previous answer from 1.

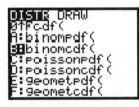   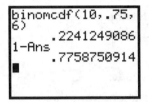

Note 10F • Sequences into Lists

With the calculator in any mode, you can use the seq(command to generate a nonrecursive sequence. To find the seq(command, press [2nd] [LIST] OPS 5:seq(.

The seq(command requires four arguments: an expression, a variable counter, the starting value of the counter, and the ending value of the counter. The counter increases in increments of 1 unless an optional fifth argument specifies a different increment.

For example, seq(X^2,X,2,6) generates the sequence of perfect squares 2^2 through 6^2. As another example, seq(X,X,11,99,2) generates the odd integers from 11 to 99. To store the sequence into a list, you can use the store key, [STO→], from the Home screen, or enter a sequence definition into the Stat Edit screen. Entering the definition in quotation marks, [ALPHA] ["], keeps the definition dynamic and allows you to edit it easily.

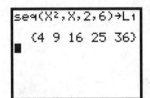

Note 11A • Entering *e*

To display the value of *e*, press 2nd [e] ENTER. To define an exponential expression or function with base *e*, press 2nd [eˣ].

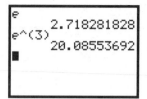

Note 11B • Normal Graphs

You can easily graph a normal curve with the normal probability distribution function, normalpdf(. To find the normalpdf(command, press 2nd [DISTR] 1:normalpdf(.

Follow these steps to graph a normal curve in Function mode:

a. Make note of the mean, μ, and the standard deviation, σ, of the distribution.

b. Press Y= and define Y₁=normalpdf(X,μ,σ). Enter the numerical values of μ and σ. Or if you have stored your data into lists and used 1-Var Stats to calculate the mean and standard deviation, you can use the exact values by pressing VARS 5:Statistics, and selecting 2:x̄ for the mean and 4:σx for the standard deviation.

c. Set an appropriate window.

d. Press GRAPH.

These screens show a normal curve with a mean 3.1 and standard deviation 0.14.

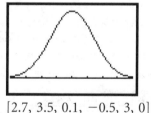

[2.7, 3.5, 0.1, −0.5, 3, 0]

To graph the standard normal distribution, that is, a normal curve with mean 0 and standard deviation 1, you need enter only normalpdf(X).

Note 11C • Probabilities of Normal Distributions

Calculating Ranges

The normal cumulative distribution function, normalcdf(, calculates the area under a normal curve between two endpoints. To find the normalcdf(command, press 2nd [DISTR] DISTR 2: normalcdf(. For a standard normal distribution with mean 0 and standard deviation 1, enter normalcdf(*lower,upper*). For any normal distribution, with mean μ

(continued)

and standard deviation σ, enter the command in the form
normalcdf(*lower,upper*, μ, σ).

 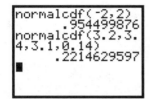

Graphing Ranges

The ShadeNorm(command graphs the normal curve and shades the area
between the specified endpoints. It also reports the probability associated
with that area. To find the ShadeNorm(command, press [2nd] [DISTR] DRAW
1:ShadeNorm(.

To use the command, first set an appropriate window. Then, on the Home
screen, enter the command in the form ShadeNorm(*lower,upper*, μ, σ).

 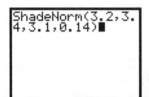 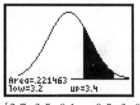

$$[2.7, 3.5, 0.1, -0.5, 3, 0]$$

Note 11D • Creating Random Probability Distributions

You can create lists of various kinds of distributions.

    a. To create a uniform distribution, use [MATH] PRB 1:rand. This example
    creates a list of 200 values uniformly distributed between 20 and 50.

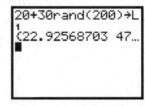 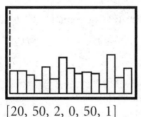

$$[20, 50, 2, 0, 50, 1]$$

    b. To create a normal distribution, use [MATH] PRB 6:randNorm(. This example
    creates a list of 200 values with mean 35 and standard deviation 5. Almost
    all of the values will be between 20 and 50.

 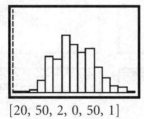

$$[20, 50, 2, 0, 50, 1]$$

(continued)

c. To create a left-skewed distribution, use the cube root of rand(. This example creates a left-skewed population of 200 values between 20 and 50.

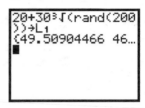 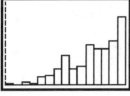

[20, 50, 2, 0, 50, 1]

d. To create a right-skewed distribution, use the cube of rand(. This example creates a right-skewed population of 200 values between 20 and 50.

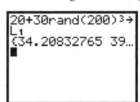 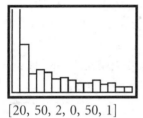

[20, 50, 2, 0, 50, 1]

Note 11E • Correlation Coefficient

There are two ways to find a correlation coefficient, *r*, using the calculator. You can manually enter the calculations yourself, or you can have the calculator do the work for you.

First store your bivariate data into two lists, say list L1 for the *x*-values and list L2 for the *y*-values.

Follow these steps to manually calculate *r*:

a. Calculate the two-variable statistics that you need for the formula by pressing [STAT] CALC 2:2-Var Stats [2nd] [L1] [,] [2nd] [L2] [ENTER].

b. Start inputting the formula $\frac{\Sigma(x - \bar{x})(y - \bar{y})}{s_x s_y (n - 1)}$ by entering sum((L1−. Do not press [ENTER] yet. To find the sum(command, press [2nd] [LIST] MATH 5:sum(.

c. Press [VARS] 5:Statistics 2:x̄ to enter x̄ into the expression. Notice that by pressing [VARS] 5:Statistics you can also get 1:n, 3:Sx, 5:ȳ, and 6:Sy.

d. Enter the rest of the formula, ((L2−ȳ))/(SxSy(n−2)).

e. Press [ENTER] to display the value of *r*.

   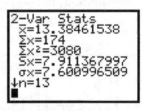

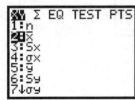 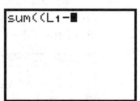 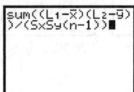 

(continued)

Follow these steps to have the calculator compute *r*:

   a. Press [2nd] [CATALOG] [D]. Scroll down to Diagnostic On. Press [ENTER] [ENTER]. (Note: You need to do this step only once. After you turn the diagnostics on, the setting remains on.)

   b. Press [STAT] CALC 8:LinReg(a+bx) [2nd] [L1] [,] [2nd] [L2] [ENTER]. (Note: You can also use 4:LinReg(ax+b) instead of 8:LinReg(a+bx).)

   c. The calculator displays the value of *r*, as well as other information about the least squares line, which you'll learn about later.

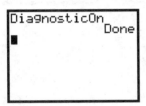 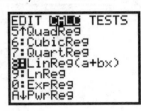  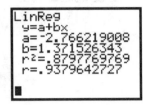

Note 11F • Least Squares Line

The calculator can find the equation of the least squares line in either the form $y = ax + b$ or the form $y = a + bx$. To find the least squares commands, press [STAT] CALC 4:LinReg(ax+b) or 8:LinReg(a+bx). Either command defaults to using list L1 for the *x*-values and list L2 for the *y*-values, but you may specify another pair of lists by following the command with the list names separated by a comma.

When you press [ENTER] the calculator displays the slope and *y*-intercept of the least squares line; the correlation coefficient, *r*; and the coefficient of determination, r^2.

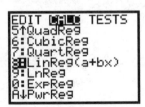 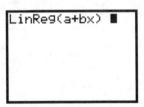 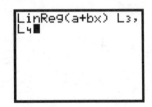 

To enter the equation of the least squares line into the Y= screen, enter a function name after the command. Find the function names by pressing [VARS] Y-VARS 1:Function.

 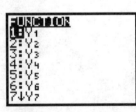 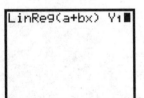

(continued)

If you forget to specify a function name, you can later paste the least squares equation into the Y= screen. Press [Y=] and go to the desired function. Then press [VARS] 5:Statistics, go to the EQ submenu, and select 1:RegEq.

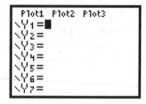  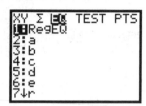 

Discovering Advanced Algebra Calculator Notes for the Texas Instruments TI-83 Plus and TI-84 Plus

Note 12A • Changing Settings

For your work in this chapter, check and change, if necessary, three settings on the Mode screen.

a. Press MODE and set the third line to Degree. You will work with angles in this chapter and those angles are measured in degrees. If you get a "funny" answer when using a trigonometric function, check to see that you are still in Degree mode.

b. Set the fourth line to Par. In the second half of this chapter you graph and use parametric equations. When you switch to Parametric mode, the Y= screen and the Window screen change.

c. Set the sixth line to Simul. Later in this chapter you may graph more than one set of parametric equations. In Simultaneous mode, all equations graph at the same time. In Sequential mode, equations graph one after the other.

Note 12B • Trigonometric and Inverse Trigonometric Functions

Before using the trigonometric functions on the calculator, press MODE to check that you are in Degree mode.

Use SIN, COS, or TAN to find the sine, cosine, or tangent ratio of any angle measure.

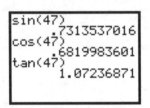

Use 2nd [SIN⁻¹], 2nd [COS⁻¹], or 2nd [TAN⁻¹] to find the angle measure that has the given ratio.

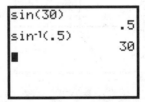
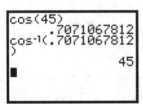
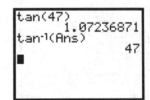

Note 12C • Graphing Parametric Equations

In Parametric mode, you define equations in terms of the parameter t. To enter the variable t, press X,T,θ,n.

It takes a pair of equations to create a single parametric graph. Until you define both X_{1T} and Y_{1T} (or any other X-Y pair) on the Y= screen, nothing will graph.

Setting the Window

In Parametric mode the Window screen is different from the familiar Function mode Window screen. The Graph screen that you see is still set by the values of Xmin, Xmax, Xscl, Ymin, Ymax, and Yscl. But in addition, you must

(continued)

set the starting and stopping values of t. The t-values you choose do not affect the dimensions of the Graph screen, but they do affect what will be drawn.

Tmin = the minimum t-value that the calculator uses to evaluate the x- and y-function values.

Tmax = the maximum t-value that the calculator uses to evaluate the x- and y-function values.

Tstep = the increment by which t increases between each evaluation. Tstep controls the speed at which the graph is drawn. Start with Tstep equal to about one-hundredth of the range of t, $\frac{\text{Tmax}-\text{Tmin}}{100}$. If the graphing speed is not to your liking or your graph needs more detail, adjust Tstep.

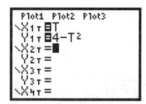  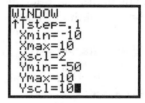 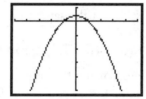

Setting the Graph Style

The graph styles are the same as those in Function mode except there is no shading in parametric equation graphs. See **Note 3A** for help with graph styles.

Note 12D • Tracing Parametric Equations

In Parametric mode, when you press [TRACE], the spider starts at the point (x, y) defined by Tmin. The t-, x-, and y-values are displayed.

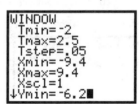 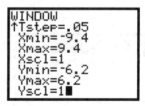 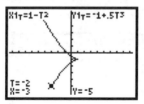

Each time you press the right arrow key, t increases by Tstep and the spider moves to the new point defined by the new t-value. Note that the right arrow key may not necessarily move the spider to the right on the graph, but it will always increase the value of t. Pressing the left arrow key similarly decreases the value of t.

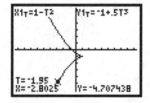 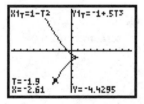

If more than one pair of equations is defined on the Y= screen, pressing the up and down arrow keys makes the spider jump to the previous or next pair

(continued)

of equations. When the spider jumps to another pair of equations, the new pair is evaluated at the current *t*-value. The spider may be anywhere on the screen depending on the *x*-value and *y*-value for the new pair of equations.

Instead of using the right or left arrow keys to increase or decrease the *t*-value, you can enter a number. The spider jumps to the point defined by that *t*-value as long as the number is between Tmin and Tmax.

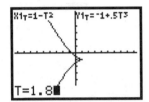 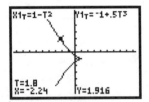

In Parametric mode there are no commands for finding the intersection of two graphs or the *x*- or *y*-intercepts of one graph. You'll need to trace in order to approximate points of intersection.

Note 12E • Parametric Walk

Use the CBRSET program to collect data for both recorder X and recorder Y. See **Note 4C** for help with the CBRSET program. For this investigation, you should collect data for 5 seconds (S) and have 100 samples (N).

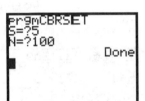

You may now disconnect the CBRs if you wish. Follow the instructions in the investigation; recorders X and Y should press their triggers on the front of their CBRs when the director says "go." When finished, reconnect the CBRs and run the CBRGET program. CBRGET stores time data in list L1 and distance data in list L2, and it will make a scatter plot of these data. In order to finish the investigation, see **Note 1F** for help with making scatter plots and see **Note 1H** for help with linking lists.

Note 12F • Graphing Functions with Parametric Equations

While in Parametric mode, the Draw menu allows you to graph a function that uses only *x* and *y*. This gives you a way to check your work when eliminating the parameter. For example, follow these steps to graph $y = 3 + 2(x - 1)$ at the same time you are displaying parametric equations:

   a. Press [2nd] [DRAW] 6:DrawF.

   b. Enter the first part of the equation, 3+2(.

   c. Press [ALPHA] [X]. You must use the alpha letter X, not [X,T,θ,n], for the variable.

   d. Enter the rest of the equation, −1).

   e. Press [ENTER]. Do not press [GRAPH] to display the graph of the function.

 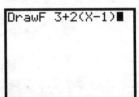 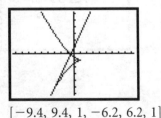

$[-9.4, 9.4, 1, -6.2, 6.2, 1]$

Note 13A • Unit Circle

Follow these steps to graph a unit circle:

a. Press MODE and set the third line to Degree and the fourth line to Par.

b. On the Y= screen, enter the equations $X_{1T}=\cos(T)$ and $Y_{1T}=\sin(T)$.

c. Set the Window screen to

$$\text{Tmin} = 0$$
$$\text{Tmax} = 900$$
$$\text{Tstep} = 15$$
$$\text{Xmin} = -2.35$$
$$\text{Xmax} = 2.35$$
$$\text{Xscl} = 1$$
$$\text{Ymin} = -1.55$$
$$\text{Ymax} = 1.55$$
$$\text{Yscl} = 1$$

d. Display the graph.

Now you can press TRACE to find the coordinates of points on the circle.

 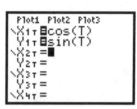 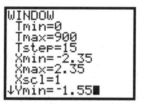

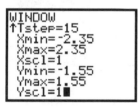 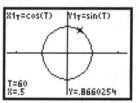

Note 13B • Radians

In Radian mode, the calculator treats the input of a sine, cosine, or tangent function as a radian measure instead of a degree measure. It also returns a radian measure when you use the inverse functions. To put the calculator in Radian mode, press MODE and set the third line to Radian.

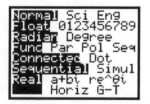 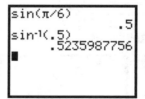

(continued)

Overriding Radian or Degree Mode

In Radian mode, enter a degree symbol, °, after the input if you want the calculator to override the Radian mode and treat the input as a degree measure. To find the degree symbol, press 2nd [ANGLE] 1:°.

 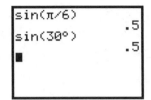

In Degree mode, enter a radian symbol, ʳ, after the input if you want the calculator to override the Degree mode and treat the input as a radian measure. To find the radian symbol, press 2nd [ANGLE] 3:ʳ.

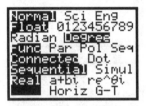  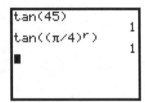

Converting between Radians and Degrees

You can use the override feature to convert an angle measure from radians to degrees or from degrees to radians.

For example, follow these steps to convert 30° to radians:

    a. Set the calculator to Radian mode.

    b. On the Home screen, enter 30 and press 2nd [ANGLE] 1:°.

    c. Press ENTER. This gives you the radian measure expressed as a decimal.

    d. To find the radian measure expressed as a multiple of π, press ÷ 2nd [π] ENTER [MATH] 1:▶Frac ENTER. Imagine that π follows the fraction. So, 30° is equivalent to $\frac{1}{6}\pi$, or $\frac{\pi}{6}$, radians.

 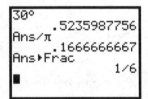

For another example, follow these steps to convert $\frac{7\pi}{12}$ radians to degrees:

    a. Set the calculator to Degree mode.

    b. On the Home screen, press (7 2nd [π] ÷ 12) 2nd [ANGLE] 3:ʳ.

    c. Press ENTER. So, $\frac{7\pi}{12}$ radians is equivalent to 105°.

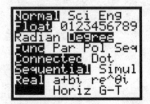 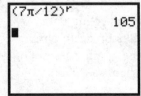

Note 13C • Secant, Cosecant, and Cotangent

The calculator does not have built-in secant, cosecant, or cotangent functions. You must calculate these functions by using the appropriate reciprocal identities.

For example, in Radian mode, to evaluate $\sec\left(\frac{\pi}{6}\right)$, press 1 $\boxed{\div}$ $\boxed{\text{COS}}$ $\boxed{\text{2nd}}$ $[\pi]$ $\boxed{\div}$ 6 $\boxed{)}$ $\boxed{\text{ENTER}}$, or press $\boxed{\text{COS}}$ $\boxed{\text{2nd}}$ $[\pi]$ $\boxed{\div}$ 6 $\boxed{)}$ $\boxed{x^{-1}}$ $\boxed{\text{ENTER}}$.

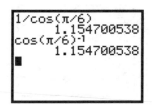

To evaluate $\csc\left(\frac{\pi}{6}\right)$, press 1 $\boxed{\div}$ $\boxed{\text{SIN}}$ $\boxed{\text{2nd}}$ $[\pi]$ $\boxed{\div}$ 6 $\boxed{)}$ $\boxed{\text{ENTER}}$, or press $\boxed{\text{SIN}}$ $\boxed{\text{2nd}}$ $[\pi]$ $\boxed{\div}$ 6 $\boxed{)}$ $\boxed{x^{-1}}$ $\boxed{\text{ENTER}}$.

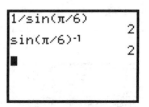

To evaluate $\cot\left(\frac{\pi}{6}\right)$, press 1 $\boxed{\div}$ $\boxed{\text{TAN}}$ $\boxed{\text{2nd}}$ $[\pi]$ $\boxed{\div}$ 6 $\boxed{)}$ $\boxed{\text{ENTER}}$, or press $\boxed{\text{TAN}}$ $\boxed{\text{2nd}}$ $[\pi]$ $\boxed{\div}$ 6 $\boxed{)}$ $\boxed{x^{-1}}$ $\boxed{\text{ENTER}}$.

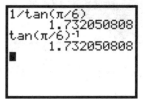

To find the inverse of a secant, cosecant, or cotangent function, use the reciprocal identity's inverse with the reciprocal of the input.

For example, in radian mode, to find $\sec^{-1}(3)$, press $\boxed{\text{2nd}}$ $[\text{COS}^{-1}]$ 1 $\boxed{\div}$ 3 $\boxed{)}$ $\boxed{\text{ENTER}}$.

To find $\csc^{-1}(3)$, press $\boxed{\text{2nd}}$ $[\text{SIN}^{-1}]$ 1 $\boxed{\div}$ 3 $\boxed{)}$ $\boxed{\text{ENTER}}$.

To find $\cot^{-1}(3)$, press $\boxed{\text{2nd}}$ $[\text{TAN}^{-1}]$ 1 $\boxed{\div}$ 3 $\boxed{)}$ $\boxed{\text{ENTER}}$.

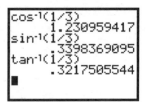

Note 13D/App • Collecting Sound Frequency Data Using the EasyData App

You need a CBL2 and the EasyData application to collect sound frequency data. Plug the microphone probe into channel CH 1 of the CBL2, and connect the calculator to the CBL2. Press $\boxed{\text{APPS}}$ and select EasyData. If the program does not recognize the microphone, follow these steps:

 a. Press Setup ($\boxed{\text{WINDOW}}$) and select Other Sensors.

 b. Select CH1 and press Next ($\boxed{\text{ZOOM}}$). Then choose Microphone and press Next.

Ring the tuning fork and press Start ($\boxed{\text{ZOOM}}$) to begin collecting data. The CBL2 collects data for only 0.02 second, so it will stop again almost immediately. The calculator will display a graph. If the graph does not look like a sinusoidal curve, press Main ($\boxed{\text{TRACE}}$) and then Start ($\boxed{\text{ZOOM}}$) to

(continued)

try again. Press OK ([GRAPH]) to overwrite the data. If you continue to have trouble collecting good data, adjust the microphone's position.

When you have good data, press Main ([TRACE]) and then Quit ([GRAPH]). The calculator will tell you that time data are stored in list L1 and sound frequency data are stored in list L2. Press OK ([GRAPH]).

Note 13E • Polar Coordinates

Graphing Polar Equations

Follow these steps to graph a polar equation:

a. Press [MODE] and set the third line to Degree and the fourth line to Pol.

b. On the Y= screen, enter a function in the form $r = f(\theta)$. Press [X,T,θ,n] to get θ.

c. On the Window screen, set values of θ as well as x and y.

d. Display the graph.

  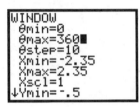 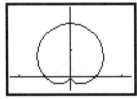

$$[-2.35, 2.35, 1, -0.5, 2.6, 1]$$

Tracing Polar Coordinates

No matter what mode you're in, you can find the polar coordinates of a point on a graph by pressing [2nd] [FORMAT] and selecting PolarGC. Then when you trace or move the cursor about the screen, you will see coordinates in the form (r, θ).

 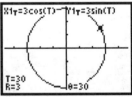

$$[-4.7, 4.7, 1, -3.1, 3.1, 1]$$

Remember to change the format back to RectCG in order to display coordinates in the form (x, y).

Comment Form

Please take a moment to provide us with feedback about this book. We are eager to read any comments or suggestions you may have. Once you've filled out this form, simply fold it along the dotted lines and drop it in the mail. We'll pay the postage. Thank you!

Your Name _____

School _____

School Address _____

City/State/Zip _____

Phone _____

Book Title _____

Please list any comments you have about this book.

Do you have any suggestions for improving the student or teacher material?

To request a catalog, or place an order, call us toll free at 800-995-MATH, or send a fax to 800-541-2242. For more information, visit Key's website at www.keypress.com.

Please detach page, fold on lines and tape edge.

NO POSTAGE
NECESSARY
IF MAILED
IN THE
UNITED STATES

BUSINESS REPLY MAIL
FIRST CLASS   PERMIT NO. 338   OAKLAND, CA

POSTAGE WILL BE PAID BY ADDRESSEE

KEY CURRICULUM PRESS
1150 65TH STREET
EMERYVILLE CA 94608-9740
ATTN: EDITORIAL

Key Curriculum Press
Innovators in Mathematics Education

Comment Form

Please take a moment to provide us with feedback about this book. We are eager to read any comments or suggestions you may have. Once you've filled out this form, simply fold it along the dotted lines and drop it in the mail. We'll pay the postage. Thank you!

Your Name _____

School _____

School Address _____

City/State/Zip _____

Phone _____

Book Title _____

Please list any comments you have about this book.

Do you have any suggestions for improving the student or teacher material?

To request a catalog, or place an order, call us toll free at 800-995-MATH, or send a fax to 800-541-2242.
For more information, visit Key's website at www.keypress.com.

Please detach page, fold on lines and tape edge.

BUSINESS REPLY MAIL
FIRST CLASS   PERMIT NO. 338   OAKLAND, CA

POSTAGE WILL BE PAID BY ADDRESSEE

KEY CURRICULUM PRESS
1150 65TH STREET
EMERYVILLE CA 94608-9740
ATTN: EDITORIAL